LAB MANUALS

Springer

Berlin
Heidelberg
New York
Barcelona
Hong Kong
London
Milan
Paris
Singapore
Tokyo

René H. Wijffels (Ed.)

Immobilized Cells

With 54 Figures and 8 Tables

Springer

Dr. RENÉ H. WIJFFELS

Wageningen University
Food and Bioprocess Engineering Group
P.O.B. 8129
6700 EV Wageningen
The Netherlands
e-mail: rene.wijffels@algemeen.pk.wag-ur.nl

ISBN 3-540-67070-X Springer-Verlag Berlin Heidelberg New York

Library of Congress Cataloging-in-Publication Data
Immobilized cells / René H. Wijffels (ed.).
  p. cm. – (Springer lab manuals)
  Include bibliographical references and index.
  ISBN 3-540-67070-X (wire bdg.: alk. paper)
  1. Immobilized cells – Laboratory manuals. I. Wijffels, René H., 1960- II. Series
  QH585.5.145 1466 2000
  571.6-dc21

Springer-Verlag Berlin Heidelberg New York
a member of BertelsmannSpringer Science + Business Media GmbH
© Springer-Verlag Berlin Heidelberg 2001

Production: PRO EDIT GmbH, 69126 Heidelberg, Germany
Cover design: design & production GmbH, 69121 Heidelberg, Germany
Typesetting: Mitterweger & Partner, 68723 Plankstadt, Germany
Printed on acid free paper   SPIN 10547923   27/3130/So   5 4 3 2 1 0

Contents

Introduction

Chapter 1
Characterization of Immobilized Cells; Introduction
RENÉ H. WIJFFELS   1

Immobilization

Chapter 2
Description of the Support Material
EMILY J.T.M. LEENEN   6

Chapter 3
Description of the Immobilization Procedures
DENIS PONCELET, CLAIRE DULIEU, and MURIEL JACQUOT   15

Subprotocol 1: Ionic Gelation (Alginate Beads).   20
Subprotocol 2: hermal Gelation (κ-Carrageenan Beads)   23
Subprotocol 3: Ionic Polymer Coating (Chitosane on Alginate Beads)   25
Subprotocol 4: Coating by Transacylation Reaction   26
Subprotocol 5: Polyelectrolyte Complex Membrane
                (Sulfoethylcellulose/Polydiallyldimethyl
                Ammonium Chloride)   27

Chapter 4
Measurement of Density, Particle Size and Shape of Support
ERIK VAN ZESSEN, JOHANNES TRAMPER, and ARJEN RINZEMA . .   31

Chapter 5
Mechanical Stability of the Support
EMILY J.T.M. LEENEN.   36

Chapter 6
Diffusion Coefficients of Metabolites
EVELIEN E. BEULING   44

Subprotocol 1: Preparation of the Membrane   48
Subprotocol 2: Diffusion Experiments   51
Subprotocol 3: Step-Response Method Utilizing Micro-Electrodes .   54

Kinetics

Chapter 7
Quantity of Biomass Immobilized, Determination
of Biomass Concentration
ELLEN A. MEIJER and RENÉ H. WIJFFELS   65

Chapter 8
Kinetics of the Suspended Cells
RENÉ H. WIJFFELS   74

Chapter 9
Diffusion Limitation
RENÉ H. WIJFFELS   77

Chapter 10
Micro-Electrodes
DIRK DE BEER   85

Subprotocol 1: Manufacturing of O_2 Microsensors   87
Subprotocol 2: Measurements   92

Chapter 11
Biomass Gradients
RENÉ H. WIJFFELS   101

Chapter 12
NMR and Immobilized Cells
JEAN-NOËL BARBOTIN, JEAN-CHARLES PORTAIS,
PAULA M. ALVES, and HELENA SANTOS 123

Engineering

Chapter 13
Immobilization at Large Scale by Dispersion
DENIS PONCELET, STEPHANE DESOBRY, ULRICH JAHNZ,
and K. VORLOP 139

Subprotocol 1: Jet Breaking Methods 140
Subprotocol 2: Lentikats 142
Subprotocol 3: Rotating Devices 144
Subprotocol 4. Emulsification Using Static Mixer 146

Chapter 14
Immobilization at Large Scale with the Resonance Nozzle Technique
JAN H. HUNIK 150

Chapter 15
External Mass Transfer
RENÉ H. WIJFFELS 162

Chapter 16
Liquid Fluidization of Gel-Bead Particles
ERIK VAN ZESSEN, JOHANNES TRAMPER, and ARJEN RINZEMA .. 175

Chapter 17
Gradients in Liquid, Gas or Solid Fractions
RENÉ H. WIJFFELS 182

Chapter 18
Support Material Stability at the Process Conditions Used
EMILY J.T.M. LEENEN 191

Cases

Chapter 19
Immobilized Cells in Food Technology:
Storage Stability and Sensitivity to Contamination
CLAUDE P. CHAMPAGNE 199

Chapter 20
Immobilized Cells in Bioremediation
BRONAGH M. HALL and AIDEN J. MC LOUGHLIN 213

Subprotocol 1: Alginate Encapsulation 214
Subprotocol 2: Carrageenan Encapsulation 217
Subprotocol 3: Co-Immobilization with Adjuncts 219
Subprotocol 4: Immobilization with Synthetic Polymers 221
Subprotocol 5: Microencapsulation 225
Subprotocol 6: Monitoring Microbial Inoculum in the Environment 230

Chapter 21
Plasmid Stability in Immobilized Cells
JEAN-NOËL BARBOTIN 235

Chapter 22
Immobilization for High-Throughput Screening
NICOLE M. NASBY, TODD C. PETERSON, and CHRISTOPHER J. SILVA 247

Subject Index 259

Characterization of Immobilized Cells; Introduction

RENÉ H. WIJFFELS

Introduction

In processes with immobilized cells, the cells are attached to or entrapped in an inert support. In a continuously operated bioreactor, medium containing the substrate will be supplied. Substrate will be converted into a product by the immobilized cells and product and remaining substrate disappear with the outflowing medium. The immobilized cells are retained easily in the bioreactor and as such utilized continuously. In this way the capacity of the process is independent of the growth rate of the micro-organisms involved. In particular in cases where cells grow slowly, immo-bilized-cell process are more advantageous over processes with suspended cells. Also for very specific situations where the presence of biomass in the product should be prevented (e.g. champagne), drugs should be added gra-dually to the medium (e.g. islets of Langerhans), phage infections in starter cultures should be prevented (e.g. lactic-acid bacteria for cheese produc-tion) and plasmids should be stabilized (for application of genetically en-gineered cells in a continuous mode), immobilized cells perform better than suspended cells.

Immobilized cells have been studied widely during the last decades. An enormous quantity of papers have been published. For many processes the complex physiology in a heterogeneous environment is now becoming clear. In addition it is shown that in many processes it is more efficient to use immobilized cells than suspended cells.

In 1995 the symposium "Immobilized Cells: Basics and Applications" was organized under auspices of the Working Party of Applied Catalysis of the European Federation of Biotechnology (Wijffels et al. 1996). The sym-

René H. Wijffels, WageningenUniversity, Food and Bioprocess Engineering Group, P.O. Box 8129, Wageningen, 6700 EV, The Netherlands (*phone* +31-317-482884; *fax* +31-317-482237; *e-mail* rene.wijffels@algemeen.pk.wag-ur.nl)

posium covered the path from basic physiological research to applications, and scientists were brought together from different disciplines from academia, industry and research institutes. For applications, physiology needs to be integrated with engineering. The goal of the symposium was to relate basic research to applications. Another aim was to extract guidelines for characterization of immobilized cells in view of process design and application from the contributions.

Laboratory methods for scientific research and manufacturers

It is important to recognize that research on immobilized metabolizing cells should lead, whenever possible, to eventual successful implementation of such catalysts in industrial processes. One of the aims of the proposed guidelines was to ensure that research done is effective in achieving this objective. It is difficult otherwise to justify the research.

Of course there are differences between scientific research and applications; applications should be fed by the scientific efforts. The scientific guidelines could be a tool to bridge the gap. As a result of following the guidelines it should not be strictly necessary for manufacturers to repeat the tests that should have been done in the scientific research phase. The guidelines if implemented should ensure a certain minimal quality of research data. Development of an application can thus focus on more specific research by manufacturers, e.g. the scale-up aspects.

The goal of research, however, is not always application oriented. Different goals of the research can be:

- development of methodology

- scientific question

- application

In the different phases of the development different parts of the guidelines should be used. If research is done in order to apply the process, the scale up and applications should be considered at a very early stage, so that experiments done, for instance, on physiology are relevant. In case of the development of methodologies and scientific questions this is not always necessary.

Guidelines should not be too complicated and be a help for research rather than a number of restrictive procedures. Guidelines should cover the whole procedure of the development of processes with immobilized cells. This means that within a single research paper not all guidelines need to be used. Generally in the development of a process, research activity will start with the physiology of the immobilized cells and be concluded by scale-up and application. Guidelines for the different phases in research are given below.

Springer Lab Manual about immobilized cells

The Springer Lab Manual about immobilized cells is structured in such a way that guidelines that resulted from the conference 'Immobilized Cells: Basics and Applications' were followed and translated into some practical methods. As for the guidelines for immobilized cells a structure was chosen in which general aspects of immobilization were covered, the kinetics were described and engineering aspects were covered. Besides laboratory methods some simple calculation procedures have been given in this lab manual. The combination of the practical methods with the calculations provide a tool for carrying out research with immobilized cells.

The lab manual concludes with some recent, promising and practical cases of immobilized cells as examples of applications with potential.

Immobilization

An important phase in studies with immobilized cells is the choice of the support material and the immobilization procedure. Materials and procedures should be compatible with the biocatalyst and the process, i.e. the immobilization procedure should be mild, diffusion of substrates and products in the support material should be possible and the support material should be stable during the process. This part covers:

• clear description of the immobilization procedure

• clear description of the support material

• density, particle size and shape of the support

• mechanical stability of the support

• diffusion coefficient of substrates and products

Kinetics

Immobilization can effect the kinetics (both intrinsically and apparently) of the biocatalyst considerably. Information about the kinetics of the immobilized cells in comparison to free cells is generally essential for application. In addition the kinetics of immobilized cells is (apparently) so complicated that detailed studies to understand the processes are necessary. Basic knowledge opens new possible applications. Methods are given to determine:

- the quantity of immobilized biocatalysts
- kinetics of the suspended cells and the effects of immobilization on the intrinsic kinetics
- whether the observed reaction rate is diffusion limited
- substrate concentration profiles
- biomass concentration profiles

Engineering

For application of processes the engineering phase is essential. In this phase previous phases are integrated and scale-up aspects are studied in detail. In this phase not only the immobilized biocatalyst will be studied but also the bioreactor:

- large scale immobilization methods
- productivity
- mass transfer description
- particles-suspension characteristics
- gradients in liquid, gas or solid fractions
- support material stability under the process conditions used

Cases

If a process appears to be feasible, application oriented research can be initiated. Research in this phase partly overlaps the engineering studies. In addition much information about the product and the process needs to be obtained which is not that specific for immobilized cells but essential for application. Some cases are described of recent (potential) applications of immobilized cells:

- in food technology

- in soil bioremediation

- of genetically modified organisms

- in screening

References

Lilly M.D. (1986) Recommendations for nomenclature to describe the metabolic behaviour of immobilized cells. Enzyme Microb. Technol. 8: 315

Wijffels R.H., Buitelaar R.M., Bucke C., Tramper J. (1996) Immobilized cells: basics and applications. Progress in Biotechnology vol. 11, Elsevier, 845 p.

Working Party on Immobilized Biocatalysts (1983) Guidelines for the characterization of immobilized biocatalysts. Enzyme Microb. Technol. 5: 304-307

Description of the Support Material

EMILY J.T.M. LEENEN

Introduction

It is not possible to describe all possible support materials, because every day new materials are being developed. In this chapter only hydrogels are considered. These gels can be divided into two groups: natural and synthetic gels. In general the support material needs to restrict the free migration of cells.

Natural (Hydro)Gels

The two most studied natural hydrogels are alginate and κ-carrageenan. Therefore these two gels are described in more detail. Other known hydrogels are (among other): agar, agarose and gelatin. Natural gels generally allow a mild immobilization procedure, which enables most cells to survive the immobilization procedure (Lewandowski et al., 1987; Tramper & Grootjen, 1986; Van Ginkel et al., 1983; Leenen et al., 1996).

Carrageenan

Carrageenan is extracted from seaweed and a gel is derived by stabilization with K^+-ions or by thermogelation (reducing the temperature at low ion concentrations). Carrageenan consists of alternating structures of D-galactose 4-sulphate and 3,6-anhydro-D-galactose 2-sulphate. The gel strength increases with the level of 3,6-anhydro-D-galactose 2-sulphate in the poly-

✉ Emily J.T.M. Leenen, National Institute of Health and the Environment, Microbiological Laboratory for Health Protection, P.O. Box 1, Bilthoven, 3720 BA, The Netherlands (*phone* +31-302473711; *fax* +31-30-2744434; *e-mail* Imke.Leenen@rivm.nl;Frank-Imke@hetnet.nl)

mer (Ainsworth & Blanshard, 1978). The carrageenan matrix becomes weak when disturbing ions are present (gelates the K^+-ions out of the matrix) or when K^+-ions are absent in the liquid phase.

Addition of gums (locust bean gum, xanthane, arabic gum: Arnaud et al., 1989; Audet et al., 1990), Al^{3+} (Chamy et al., 1990) or coating with a polymer (synthetic gel) can improve the gel characteristics (Chapter 3).

General characteristics (Leenen et al., 1996):

- cells survive the mild immobilization methods with carrageenan

- cells grow easily in the matrix

- the diffusion coefficients of substrates are high

- relatively cheap

- the matrixes are soluble

- biodegradable

- relatively weak

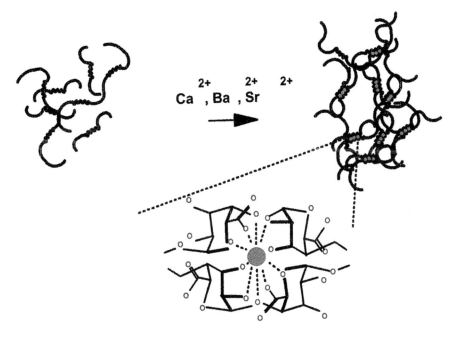

L-guluronic acid -(1-4)- L-guluronic acid

Fig. 1. The formation of an alginate gel by binding with divalent cations

Alginate

Alginate is derived from algae and can be stabilized by a divalent cation. It consists of 1-4 bonded D-mannuronic- and L-guluronic-acid groups (Figure 1).

Gels are formed due to binding of divalent cations to the guluronic acid groups. The strength of the gel depends strongly on the origin of the alginate (type of algae, stem, leave; Martinsen et al., 1989). The quality and strength of the matrix improves with an increasing amount of guluronic acid groups in the alginate (Figure 2; Smidsrød & Skjåk-Bræk, 1990). The composition of some alginates is given in Table 1.

Alginate originating from the stem of *Laminarea hyperborea* has a high concentration of guluronic acid. Most alginates generally used are much weaker; they are mostly from *Macrocystis pyrifera* or *Ascophyllum nodosum*.

The type of divalent cation used also regulates the gel strength. The strength of an alginate gel increases with the affinity of the used cation according to the following order:

$$Pb^{2+} > Cu^{2+} = Ba^{2+} > Sr^{2+} > Cd^{2+} > Ca^{2+} > Ni^{2+} > Zn^{2+} > Co^{2+}$$

A few of these cations are toxic and therefore not applicable in some applications. Mostly, Ca^{2+}, Ba^{2+} and Sr^{2+} or a combination of these cations

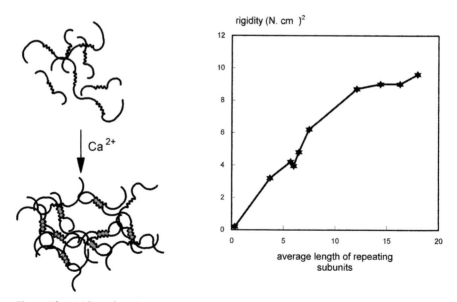

Fig. 2. The rigidity of an alginate gel as a function of the length of the guluronic acid blocks

Table 1. Source, composition and average length of guluronic acid blocks of several commercial available alginates (Smidsrød & Skjåk-Bræk, 1990).

Source	Molecular fraction	Average length of guluronic acid blocks
Laminaria hyperborea (outer layer)	0.75	17.5
Laminaria hyperborea (stem)	0.70	15.0
Laminaria hyperborea (leave)	0.55	7.3
Laminaria digitata (whole plant)	0.41	6.0
Macrocystis pyrifera	0.43	6.3
Ascophyllum nodosum (young tissue)	0.10	3.0
Ascophyllum nodosum (old tissue)	0.36	3.9
Durvillea antarctica (whole plant)	0.34	5.3

are used. Alginate gels are comparable with carrageenan gels in respect to their solvability in media with chelating chemicals, like phosphate, citrate or lactate. These compounds can extract the divalent cation used for gel formation from the matrix with a weaker gel as a result. When this happens the hydrogel swells first after which it dissolves completely.

General characteristics (Leenen et al., 1996):

- cells survive the mild immobilization methods
- cells grow easily in the matrix
- the diffusion coefficients of substrates are high
- relatively cheap
- the matrixes are soluble
- biodegradable
- relatively weak
- strength depends on the amount of guluronic acid groups
- strength depends on divalent cation used

Other natural hydrogels

Agar and agarose gels are made by thermo-gelation. The gels are made by dispersing the appropriate amount of agar/agarose in demineralized water followed by heating to 50-60°C to dissolve the powder. After cooling the gels are ready.

Those obtained are relatively weak, but can be improved by crosslinking with glutaraldehyde. This, however, decreases the survival rate of the immobilized cells (Mignot & Junter, 1990). Gelatin gels are also mostly crosslinked with glutaraldehyde or polyaldehydes. The use of dextran or starch in particular can improve the matrix drastically (Parascandola et al., 1990).

General characteristics:

- cells can survive the immobilization methods

- cells grow easily in the matrix

- the diffusion coefficients of substrates are high

- the matrixes are soluble

- biodegradable

- relatively weak

Synthetic gels

Lately, several gel-forming synthetic prepolymers have been developed. Generally gels are formed by polymerization or crosslinking. This polymerization or crosslinking is, however, done in the presence of the micro-organisms to be immobilized. This environment can be rather hostile for these cells. Activity losses of more than 90% have been reported for polyacrylamide and epoxy resins (Sumino et al., 1992; Tanaka et al., 1991). Only the milder procedures and techniques will be discussed here.

General characteristics of synthetic gels (Leenen et al., 1996):

- low or no solubility

- low or no biodegradability

- strong

- diffusivity of substrates relatively low

- recovery of immobilized cells relatively low

- relatively expensive

Polyvinyl alcohol (PVA)

PVA gels can be formed by:

- crosslinking with boric acid (Hashimoto & Furukawa, 1987; Wu & Wisecarver, 1992)

- iterative freezing and thawing (Ariga et al., 1987; Asano et al., 1992; Myoga et al., 1991)

- phosphorilization (Chen & Liu, 1994)

- photocrosslinking (Ichijo et al., 1990)

Most of these reactions are harsh, which results in a relatively low recovery of the immobilized cells in comparison with natural gels.

The mildest immobilization method is the freezing-thawing method. The survival of cells can be improved by adding cryoprotectants (e.g. disaccharides, polyalcohols) to the cell-polymer suspension before freezing. Another important factor is the freezing and thawing rates used with this method. The highest survival rates are generally found with a high freezing rate (instantaneous) and a low thawing rate (Leenen, 1997).

Polyurethane and polyethylene glycol (PEG)

Polyurethane is a physical and mechanical stable gel, but the monomers are toxic. Activity losses of 60-90 % are the result of this (Sumino et al., 1992). Vorlop and coworkers (Vorlop et al., 1992) developed a less toxic poly(carbamoylsulphonate) gel (PCS) by blocking the toxic groups during the polymerization reaction. Survival rates of 99% were reported by them (Willke et al., 1994). In general polyurethane gels are strong and not biodegradable, but attachment of other cells at the surface has been reported (Pascik, 1989; Leenen, 1997).

PEG gels have been developed and used in wastewater treatment systems by Tanaka and coworkers (1991; Sumino et al., 1992[a]). The gels have shown high stability in wastewater treatment systems. The Holland Biomaterials Group (the Netherlands) developed a similar gel, which is also physically and chemically stable.

Suppliers

Carrageenan

Copenhagen Pectin A/S
16, Ved Banen
4623 Lille Skensved
Denmark (Genugel)
fax: 45-53669446

Several chemical companies

Alginate

Kelco, Int.
Waterfield
KT20 5HQ Tadworth Surrey
UK
Fax: 44-1-737377100

Pronova Biopolymer
P.O. Box 494
3002 Drammen
Norway
Fax: +47-32203510.

Sigma
Several chemical companies

Other natural hydrogels

Several chemical companies

Polyvinyl alcohol (PVA)

Sigma (USA)
Several other chemical companies

Polyurethane and polyethylene glycol (PEG)

Federal Agricultural Research Centre
Braunschweig
Germany (group of Vorlop)

Hitachi Plant Co.
Japan

Degremont
France

Holland Biomaterials Group
Enschede
The Netherlands

Several chemical companies

References

Ainsworth PA, Blanshard JMV (1978) The interdependence of molecular structure and strength of carrageenan/carob gels. Part 1, Lebensm.-Wiss. U-Technol., 11: 279-282
Ariga O, Takagi H, Nishizawa H, Sano Y (1987) Immobilization of microorganisms with PVA hardened by iterative freezing and thawing, J. Ferment. Technol. 65: 651-658.
Arnaud JP, Lacroix C, Choplin L (1989) Effect of lactic acid fermentationon the rheological properties of (-carrageenan/locust bean gum mixed gels inoculated with S. thermophilus, Biotechnol. Bioeng. 34: 1403-1408.
Asano H, Myoga H, Asano M, Toyoa M (1992) Nitrification treatability of whole microorganisms immobilized by the PVA freezing method
Audet P, Paquin C, Lacroix C (1990) Batch fermentations with a mixed culture of lactic acid bacteria immobilized separately in (-carrageenan/locust bean gum gel beads, Appl. Microbiol. Biotechnol. 32: 662-668
Chamy R, Nunez ML, Lema JM (1990) Optimization of the hardening treatment of S. cerevisae bioparticles, Biotechnol. Bioeng. 30:52-59.
Chen K-C, Liu Y-F (1994) Immobilization of microorganisms with phosphorylated polyvinyl alcohol (PVA) gel, Enzyme Microb. Technol. 16: 79-83.
Hashimoto S, Furukawa K (1987) Immobilization of activated sludge by PVA-boric acid method, Biotechnol. Bioeng. 15: 52-59.
Ichijo H, Nagasawa J, Yamauchi (1990) Immobilization of biocatalysts with poly(vinyl alcohol) supports, J. Biotech. 14: 169-178
Leenen EJTM, Martins dos Santos VAP, Grolle KCF, Tramper J, Wijffels RH (1996) Characteristics of and selection criteria for support materials for cell immobilization in wastewater treatment, Water Res. 30: 2895-2996.

Leenen EJTM (1997) Nitrification by immobilized cells, Model and practical system. Dissertation Agricultural University Wageningen, the Netherlands.

Lewandowski Z, Bakke C, Characklis WG (1987) Nitrification and autotrophic denitrification in calcium alginate beads, Wat. Sci. Tech. 19: 175-182.

Martinsen A, Skjåk-BræK G, Smidsrød O (1989 Alginate as immobilization material: I. Correlation between chemical and physical properties of alginate gel beads, Biotechnol. Bioeng. 33: 79-89.

Mignot L, Junter G-A (1990) Diffusion of immobilized-cell agar layers: influence of bacterial growth on the diffusivity of potassium chloride, Appl. Microbiol. Biotechnol. 33: 167-171

Myoga H, Asano H, Nomura Y, Yoshida H (1991) Effects of immobilization conditions on the nitrification treatability of entrapped cell reactors using PVA freezing method, Wat. Sci. Tech. 23: 1117-1124.

Parascandola P, Alteriis E de, Scardi V (1990) Immobilization of viable yeast cells within polyaldehyde-hardened gelatin gel, Biotech. Techn. 4: 237-242.

Pascik I (1989) Modified polyurethane carriers for the biochemical waste water treatment. In: Technical advances in biofilm reactors, Nice: 49-58.

Smidsrød O, Skjåk-Bræk G (1990) Molecular structure and physical behavior of seaweed colloids as compared with microbial polysaccharides. In: Seaweed resources in Europe: Uses and potential (Giury MD, Blunden G. eds.), 185-217

Sumino T, Nakamura H, Mori I, Kawaguchi Y, Tada M (1992) Immobilization of nitrifying bacteria in porous pellets of urethane gel for removal of ammonium nitrogen from wastewater, Appl. Micrbiol. Biotechnol. 36: 556-560.

Sumino T, Nakamura H. Mori N, Kawaguchi Y (1992a) Immobilization of nitrifying bacteria by polyethylene glycol prepolymer, J. Ferment. Bioeng. 73: 37-42.

Tanaka K, Tada M, Kimata T, Harada S, Fujii Y, Mizuguchi T, Mori N, Emori H (1991) Development of new nitrogen removal systems using nitrifying bacteria immobilized in synthetic resin pellets, Wat. Sci. Tech. 23: 681-690.

Tramper J, Grootjen DRJ (1986) Operating performance of *Nitrobacter agilis* immobilized in carrageenan, Enzyme Microb. Technol. 8: 477-480

Van Ginkel CG, Tramper J, Luyben KChAM, Klapwijk A (1983) Characterization of *Nitrosomonas europaea* immobilized in calcium alginate, Enzyme Microb. Technol. 5: 297-303.

Vorlop K-D, Muscat A, Beyersdorff J (1992) Entrapment of microbial cells within polyurethane hydrogel beads with the advantage of low toxicity, Biotechnol. Tech.: 483-488.

Willke B, Willke T, Vorlop K-D (1994) Poly(carbomoyl sulphonate) as a matrix for whole cell immobilization - Biological characterization, Biotechnol. Techn. 8: 623-626.

Wu K-YA, Wisecarver KD (1992) Cell immobilization using PVA crosslinked with boric acid, Biotechnol. Bioeng. 39: 447-449.

Description of the Immobilisation Procedures

DENIS PONCELET, CLAIRE DULIEU, and MURIEL JACQUOT

Introduction

Describing all the different methods of cell immobilisation would require a complete encyclopaedia. An exhaustive revue was however published four years ago [Willaert, 1996] and may be used as reference. In the present contribution, the most common techniques of microencapsulation will be described. The reader may easily extrapolate most other methods too.

Encapsulation of cells is based on two main steps: droplet formation followed by solidification (gelation, membrane formation). In a first part, methods required to form droplets will be described using simple drawings and equations. Most systems may be easily fabricated, using standard commercial items and small set-up. In the second part, bioencapsulation methods themselves will be described.

Droplet formation methods (Lab Scale)

The most simple method to form a droplet is to extrude a liquid through a nozzle using a syringe connected to a needle. However, it results in large beads (2 to 4 mm). The following advice may be applied to all devices:

- Continuous liquid flows may be obtained using a syringe pump (most lab suppliers) or by applying air pressure on the syringe or a vessel (EFD Inc., USA, see Figure 1a). The syringe pump allows to set up directly the

✉ Denis Poncelet, ENITIAA, Rue de la Géraudiere, BP 82 225, Nantes, 44 322, France (*phone* +33-251-785425; *fax* +33-251-785467; *e-mail* poncelet@enitiaa-nantes.fr; *homepage* BRG.enitiaa-nantes.fr)
Claire Dulieu
Muriel Jacquot

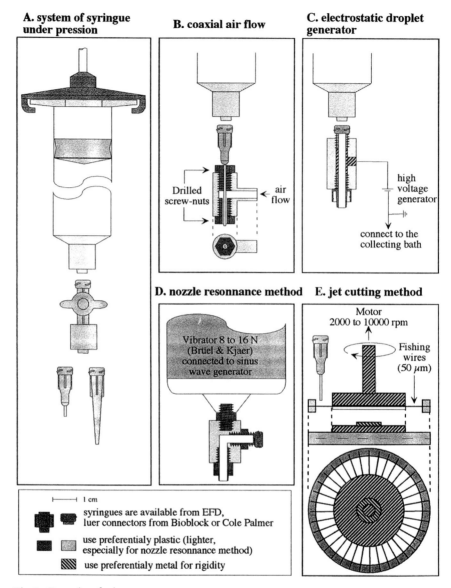

A. system of syringue under pression

B. coaxial air flow

Drilled screw-nuts

air flow

C. electrostatic droplet generator

high voltage generator

connect to the collecting bath

D. nozzle resonnance method

Vibrator 8 to 16 N (Brüel & Kjaer) connected to sinus wave generator

E. jet cutting method

Motor 2000 to 10000 rpm

Fishing wires (50 μm)

1 cm

syringues are available from EFD, luer connectors from Bioblock or Cole Palmer

use preferentialy plastic (lighter, especially for nozzle resonnance method)

use preferentialy metal for rigidity

Fig. 1. Dropping devices

flow rate. An application of pressure on the syringe permits extrusion of fluids with higher viscosities and higher flow rates.

- Pressure required to obtain a desired flow rate is mainly proportional to the length of the needle and to the inverse of the internal nozzle diameter. It may be best to select the shortest needle cut at 90 °C (Hamilton, Switzerland or EFD Inc., USA) or, preferably, a tapered injector (EFD Inc., USA). Other injectors (such as those used for printer nozzles) may be more efficient (shorter length) but more difficult to set up.

- By using shorter and larger tubing connections and low resistance injectors, the necessary pressure may be decreased by a factor up to 5, allowing the handling of highly viscous liquid.

Different systems may be adapted to the nozzle to obtain smaller droplets, low size dispersions and larger flow rates (Figure 1b to d). Two working regimes exist to form droplets. At low flow rates ($\approx$ 30 ml/h), liquid exits the nozzle as droplets. For larger flow rates, the liquid forms a jet which is broken into smaller droplets.

Within the droplet formation dropping regime, smaller droplets may be obtained by applying a coaxial air flow around the needle (Figure 1b). While increasing the air flow rate, smaller droplets are obtained but it leads quickly to droplet size dispersion and spray dispersion [Poncelet, 1993]. This method is only recommended for low flow rate and droplet size larger than 800 µm.

An alternative method involves application of an electrostatic potential between the needle and the collecting solution (Figure 1c). The droplet diameter is mainly defined by the following equation [Poncelet, 1998]:

$$d^3 = d_0^3 \left(1 - U^2/U_c^2\right) \tag{1}$$

where d is the bead diameter, do is the beads diameter without electrostatic potential (2 mm), U is the applied potential and Uc the critical potential ($\approx$ 4 kV). When U approaches Uc, the liquid forms a jet which breaks into droplet of 150 to 250 µm (independently of the potential electrostatic value). Narrow size distribution (less than 15 %) is obtained. Until now, experiments have only been conducted for low flow rate (30 ml/h). No commercial system is available.

In the jet regime, the nozzle resonance method (Figure 1d) involves application of a vibration to the liquid or nozzle. This method will be more fully described in Chapter 14. As a very simple approach, the size of the droplets is equal to 1.8 times the internal nozzle diameter and the size distribution limited to 5 %. The maximum flow rate is approximately equal to 3 l/h times the cube of the droplet diameter (in mm), allowing very high flow rate for large droplets. The jet may not break into small droplets (less than 800 µm) for high viscous liquid (more 0,5 Pa · sec) due to damping effects. Assembly set-up must avoid any external vibration. Several companies have proposed such systems (Inotech, Switzerland; Sodeva, France; Microdrop, Germany; Institute für Mikrotechnik Mainz, Germany).

Another simple method is to cut the jet using a vaned "wheel" (Figure 1e) [Pruesse, 1998]. This device is simple to build and operate and its results in similar droplet size dispersion and flow rate as the resonance nozzle method. However, this technique may be less limited in regard to the solution viscosity than the resonance nozzle method. For droplet sizes lower than 500 µm, building of the device may be more difficult requiring a very high speed motor (10 000 to 40 000 rpm). As this method is useful for high flow, it will be more fully described in Chapter 13. geniaLab BioTechnologie (Germany) has proposed a commercial form of this device.

Emulsification methods (Lab Scale)

At the laboratory scale, emulsification is performed as a batch operation (Figure 2). A continuous system will be described in Chapter 13. It may be difficult to extrapolate directly from lab scale to pilot and industrial scales. It is necessary to respect some guidelines in building this device (see Figure 2 for an operational design):

- For expensive constituents, use not less than 100 ml beaker but larger volume (500 ml or even more) is preferable.

- In many cases, encapsulation involves polymers that will accumulate in every dead (unmixed) zones. Preference should be given to a round bottom reactor (Figure 2).

- Baffles (see Figure 2) enhance the effect of the mixing and allow the establishment of a relation between droplet size and impeller speed.

- Liquid height in the container must be approximately equal to reactor diameter.

- Turbine is the most efficient impeller. Marine would work at small scale but is more efficient at pumping liquid than creating shear breakage.

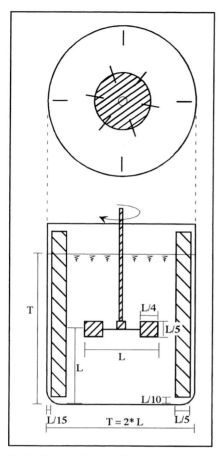

Fig. 2. Design for a turbine reactor

- Dispersed phase fraction may range from 15 to 25 % of the liquid volume. Higher values lead to larger size dispersion.

- As a first estimate, equations [Wang, 1986] may be used to compute the mean droplet diameter (50 to 500 µm).

$$\frac{D_{32}}{L} = 0.053\,We^{3/5} \tag{2}$$

with $We = \rho_c N^2 L^3/\sigma$ and $D_{32} = \frac{\Re_i n d_i^3}{\Re_i n d_i^2}$ where D_{32} is the slauther or the mean surface diameter, L the turbine diameter, We, the Weber's number, ρ_C the density of the continuous phase, N the turbine rotational speed, σ the interfacial tension and d_i the individual drop diameter.

- Size distribution follows mainly a log normal law. But for simplicity, most authors have described it as a normal law with standard deviation generally ranging between 30 % and 50 %.

- To ensure equilibrium of the dispersion, mixing must be maintained for 5 min (low viscosity) to 15 min (high viscosity, polymer solutions).

- Addition of emulsifiers reduces diameters and size dispersion. For water in oil dispersion, Span 85 at 1 to 2 % may be used.

- Separation and washing of the microcapsules from oil may be a tedious problem, especially for small sizes. In most simple situations, after "solidification" of the droplets, microcapsules are filtered on buchner with a 50 μm nylon mesh and washed with a spray of water or adequate buffer.

- Rapid solidification process (at least at droplet surface) is necessary to get individual microcapsules and narrow size dispersion.

Subprotocol 1
Ionic Gelation (Alginate Beads)

Principle   Alginate solution gels in presence of most di- and trivalent cations such as calcium, barium or aluminium. The simplest method to form alginate beads involves droplets extrusion of alginate solution into calcium solution (method 1: extrusion/external gelation) [Kierstan, 1977]. Another method involves dispersing alginate in oil phase and then releasing calcium from inside the droplets by gentle acidification (method 2: emulsification/internal gelation) [Poncelet, 1995].

Materials

- Droplet formation (method 1) or emulsification (method 2) device
- beakers, magnetic stirrer, 50 μm nylon mesh filtration device
- 100 mM calcium chloride solution
- 2 % alginate solution (Kelco, UK; SWK, France; Pronova, Norway):
  - In a beaker introduce 100 ml deionised water.
  - Mix gently with a magnetic stirrer.

- Slowly disperse 2 g alginate powder on the liquid surface.
- Leave to stand for two hours to allow alginate grain swelling.
- Mix strongly for 15 min.
- Adjust pH to 7.5 to 8 (method 2).
- Leave to stand overnight to deaerate

- freshly prepared very fine 10 % calcium carbonate (2 μm grains, Seta-carb from Omya, France) suspension in deionised water (method 2)

- canola oil mixed with 1.5 % span 80 (method 2)

- 0.5 ml acetic acid in 10 ml canola oil (method 2)

▨▨ Procedure

Method 1: Droplet formation & external gelation

1. Introduce 20 ml alginate into the droplet formation device.

2. Introduce 100 ml calcium chloride into a beaker.

3. Mix with a magnetic stirrer at low speed.

4. Extrude the alginate solution dropwise into the calcium bath.

5. Leave to stand for 30 min to insure full gelation.

6. Filter the bead on a 50 μm nylon mesh and rinse with water.

7. Hold beads in calcium chloride solution.

Method 2: Emulsification & internal gelation

1. Introduce 100 ml oil phase into the emulsifying device.

2. Homogeneously disperse 2 ml of calcium carbonate suspension into 20 ml alginate solution.

3. Start the mixing and add the alginate mixture to the oil phase.

4. Let mix for 15 min, then add the acetic acid/oil solution.

5. Allow to gelify 10 min.

6. Filter on 50 μm nylon mesh and wash with calcium chloride solution.

Results

As alginate droplets shrink during gelation, approximately 10 g of spherical alginate beads may be obtained. Diameter and size distribution is a function of the dispersion device.

Comments

- Further information about this method may be found in Martisen 1989.

- Material to be encapsulated is introduced to the alginate solution. Procedure must then be adapted to keep around 2 % final alginate concentration.

- Bead strength is a function of the guluronic contents more than the viscosity or molecular weigh. Low viscous alginate solutions facilitate the handling and extrusion.

- Higher strength may be obtained by using barium, aluminium or higher concentration calcium solutions. Increasing concentration of the alginate solution may also enhance the bead strength but hinder the solution handling due to high viscosity.

- Internal gelation and for external gelation, with high calcium concentrations or addition of sodium ions in the gelation bath lead to homogeneous alginate concentration through the bead.

For method 2 (emulsification/internal gelation)

- Small calcium carbonate grain size ($\approx 2 \, \mu m$) and good dispersion of the grains in the alginate solution are important to achieve good internal gelation.

- Initial alginate mixture pH must be higher than 7.5 to avoid internal pre-gelation.

- Complete gelation is reached at pH 6.5. Alginate mixture may be buffered to limit pH drop. But, quantity of added acetic acid must be kept constant (to maintain transfer rate).

Subprotocol 2
Thermal Gelation (κ-Carrageenan Beads)

Thermal gelation is based on the gelification of polymer solution by temperature decrease. κ-carrageenan beads may be obtained by dropping hot κ-carrageenan solution from a temperature controlled droplet extension device into a cold KCl solution (method 1: extrusion). An alternative method consists in dispersing the warm κ-carrageenan solution into vegetal oil and then dropping the temperature by adding cold oil (method 2: emulsification) [Audet, 1989].

Principle

Materials

– Droplet formation (method 1) or emulsifying (method 2) device with temperature control

– beakers, magnetic hot plate stirrer, 50 μm nylon mesh filtration device

– 200 ml vegetal oil (method 2)

– 200 ml KCl 0.3 M

– 3 % κ-carrageenan solution (Kelco, UK; SWK, France; Copenhagen Pectin, Denmark)
  – Heat 100 ml of deionised water at 80 °C in beaker.
  – Maintaining the temperature and mixing with a magnetic stirrer, add 3 g of κ-carrageenan.
  – Leave to stand until complete dissolution.
  – Reduce the temperature (to 45 °C for direct use).

Procedure

Method 1: Dropping method

1. Heat 20 ml of 2 % κ-carrageenan solution to 45 °C.

2. Introduce it into a temperature controlled dropping device.

3. Place 100 ml of cold (10 °C) KCl solution in a beaker.

4. Drop the κ-carrageenan solution in the KCl solution under gentle magnetic stirrer agitation.

5. Leave to stand for 15 min before collecting beads and keep in KCl solution.

Method 2: Emulsification method

1. Heat 100 ml of vegetal oil and 20 ml of 2 % κ-carrageenan solution to 45 °C.

2. Cool 100 ml of oil and 200 ml KCl solution to near 4 °C.

3. Place the warm oil in a temperature controlled emulsifying device.

4. Under agitation, add 20 ml of κ-carrageenan solution.

5. After 15 min, add 100 ml cold oil to the reactor.

6. After 5 min, pour the dispersion into 200 ml of cold KCl solution.

7. Let beads transfer in aqueous phase before removing oil.

8. Collect beads on a 50 μm nylon mesh and keep in KCl solution.

Results

Both procedures lead to around 20 g of κ-carrageenan beads.

Comments

- The main drawback of κ-carrageenan is the need to introduce the encapsulated material in a hot solution (45 °C). However, Copenhagen Pectin (Denmark) proposes κ-carrageenan with a low melting point (28 °C).

- Different additives have been proposed to enhance κ-carrageenan bead strength such as 10 % Locust bean gum [Arnaud, 1989] (SKW, France; Kelco, UK) or Konjac (FMC, USA).

- The method may be adapted for 2 % agarose or gellan gum in cold water (Kelco, UK) [Norton, 1990].

Subprotocol 3
Ionic Polymer Coating (Chitosane on Alginate Beads)

Polyanionic beads may be coated by polycationic membranes by simply suspending beads in the polycationic solution. This first coating may be followed by a second coating with a polyanionic polymer, and so on.   **Principle**

Materials

- beakers, magnetic hot plate stirrer, 50 μm nylon mesh filtration device
- 10 g of alginate beads
- 100 ml chitosane solution 0.2 % (Pronova, Norway; Primex, Norway)
  - Introduce 1 % acetic acid solution into a beaker.
  - Under agitation slowly, add 0.2 g of chitosane and let stand until dissolution.
  - Raise the pH to 6.
- 100 ml alginate solution 0.2 % (dilute from 2 % solution, see above for preparation)

Procedure

1. Introduce 100 ml of chitosane solution in a beaker.

2. Under gentle agitation, add 10 g of alginate beads.

3. After 30 min, filter and rinse with deionised water.

4. If required repeat 1 to 3 with alginate solution 0,1% then chitosane again, and so on.

Results

Coating leads often to reduction of the internal polymer bounds and then to swelling. 15 to 20 g of coated beads may be obtained from 10 g of alginate beads.

Comments

- Useful information about this method may be found in [Gaserod, 1998].

- This procedure has been mainly used with alginate beads coated but poly-L-lysine, polyethyleneimine and poly-ornithine have been tested as coating material.

- Multicoating allows a better control of the membrane molecular cut-off but may reduce mass transfer.

- Final polyanionic coating is preferred for biocompatibility in transplantation.

- Lower molecular weight of the coating material leads to heavier coating.

- In case of coated alginate beads, the alginate may be dissolved by suspending beads in 10 % sodium citrate.

Subprotocol 4
Coating by Transacylation Reaction

Principle   By suspending hydrogel beads containing a polysaccharidic ester and a polyamine in an alkaline solution, transacylation between these components leads to a membrane formation at bead periphery [Lévy, 1996].

Materials

- beakers, magnetic plate stirrer, 50 μm nylon mesh filtration device

- 5 g of alginate beads prepared by droplet formation methods (see above) with a solution of 1 % alginate, 2 % propylene glycol alginate (PGA) and 5 % human serum albumin (HSA).

- NaOH 1 M and HCl 1 M

Procedure

1. Introduce 50 ml of deionised water and 5 g of alginate beads in a beaker.

2. Under gentle agitation, add 800 μl of NaOH 1 M.

3. After 15 min, neutralise with HCl 1M.

4. After 15 min, filter beads and rinse with deionised water.

5. Internal bead may be dissolved by suspension in 50 ml of 10 % sodium citrate.

▪▪ Comments

- Coating thickness is controlled by the residence time in the alkaline buffer.

- pH remains neutral inside the beads. Its rises only in the microenvironement of the forming membrane.

- PGA may be replaced by pectin (less strong membrane) and HSA by several proteins such ovalbumin, haemoglobin (or most probably other polyamine materials).

Subprotocol 5
Polyelectrolyte Complex Membrane (Sulfoethylcellulose/Polydiallyl-dimethyl Ammonium Chloride)

By dropping a polyanionic solution in a polycationic solution, it forms a membrane around the droplets by polymer ionic interactions (and precipitation or coacervation) resulting in stable microcapsules [Dautzenberg 1996].

Principle

▪▪ Materials

- droplet formation device

- beakers, magnetic stirrer, 50 µm nylon mesh filtration device

- 20 ml of 3 % sodium sulfoethyl cellulose (SSEC, Wolf Walsrode, Germany)

- 100 ml Polydiallyldimethyl ammonium chloride solution 0.5 % (PDAD-MAC, Wolf Walsrode, Germany; Clariant, Germany)

- 100 ml of 0,9 % NaCl solution

Procedure

1. Introduce 100 ml of PDADMAC solution in a beaker.

2. Introduce 20 ml of SSEC solution into the droplet formation.

3. Under gentle agitation, drop the SSEC solution into PDADMAC solution.

4. Leave to stand under agitation for 30 min.

5. Filter capsules and rinse with water.

6. Stock them in saline solution.

Comments

- Capsules have thin, strong and well defined porosity (molecular cut-off as low as 3000 Daltons).

- Viscosity of the droplet solution must be higher than collecting solution.

- Many material may be used to form such capsules [Hunkeler, 1997] but strong charges at least on one of the polymers are required to get good capsules. Excess charges may lead to capsule shrinking.

- Properties of the membrane are very dependent on polymer characteristics such as polymer molecular weight and its distribution.

References

Arnaud J. P., Choplin L. et al. (1989) Rheological Behaviour of Kappa-Carrageenan/ Locust Bean Gum Mixed Gels. J Texture Stud 19: 419-430

Audet P., Lacroix C. (1989) Two-Phase Dispersion Process for the Production of Biopolymer Gel Beads: Effect of Various Parameters on Bead Size and their Distribution. Proc Biochem 24: 217-225

Dautzenberg H., Arnold G., et al. (1996) Polyelectrolyte complex formation at interface of solutions. Progr. Colloid Polym. Sci. 101: 149-156

Gaserod O. (1998) Microcapsules of Alginate-chitosan: a study of capsules formation and functional properties. PhD Thesis, NTNU, Throndheim, Norway

Hunkeler D. (1997) Polymers for Bioartificial Organs. Trends in Polym. Sci. 5:286-292

Kierstan M., Bucke C. (1977) The Immobilization of Microbial Cells, Subcellular Organelles, and Enzymes in Calcium Alginate Gels. Biotechnol Bioeng 19: 387-397

Lévy M.-C., Edwards-Lévy F. (1996) Coating alginate beads with cross-linked biopolymers: a novel method based on a transcylation reaction. J. Microencapsul. 13: 169-184

Martisen A (1989) Alginates as immobilization materials, PhD Thesis, NTH, Trondheim, Norway

Norton, S., Lacroix C. (1990) Gellan Gum Gel as Entrapment Matrix for High Temperature Fermentation Processes: a Rhelogical Study. Biotechnol Technique 4(5): 351-356

Poncelet D., Poncelet De Smet B., Beaulieu C., Neufeld R.J. (1993) Scale-up of gel bead and microcapsule production in cell immobilization. in: Fundamentals of annimal cell encapsulation and immobilization. CRC Press, Boca Raton, USA, pp 113-142

Poncelet D., Poncelet De Smet B., et al. (1995). Production of alginate Beads by Emulsification/Internal Gelation II: physicochemistry.Appl Microbiol Biotechnol 42: 644-650

Poncelet D., Neufeld R.J., et al. (1999) Formation of Microgel beads by electric dispersion of polymer solution. AIChE J. 45(9) 2018-2023

Prusse U., Fox B., et al. (1998) The jet cutting Method as a new immobilization technique. Biotechnol. Techn. 12, 105-108

Wang C.Y., Calabresse R.V. (1986) Drop Breakup in turbulent stirred-tank contactors. AIChE J. 32: 657-676

Willaert R.G, Baron G.V. (1996) Gel entrapment and microencapsulation: methods, applications and engineering principles. Reviews Chem. Eng.12: 1-205

Suppliers

Bruel & Kjaer, 2850 Naeren, Danemark (Fax: 45 42 80 14 05)

Bioblock Sci., BP 111, 67 403 Illkirch cedex (Fax: 33-3-88 67 11 68)

TEXT fehlt, Division Surfactants, 84504 Burgkirchen, Germany
(Fax: 49-8679 75063)

Cole-Palmer, 625 EastBunker Court, Vernon Hill, Il 60061-1844 USA
(Fax: 1-847-549 7600)

Copenhagen Pectin A/S, 16, Ved Banen, 4623 Lille Skensved, Denmark
(Fax: 45-53 66 94 46)

EFD, 977 Waterman Ave., East Providence RI 02914 USA
(Fax: 1-401-431 0237)

FMC Corp, 2000 Market Street, Philadelphia, PA 19103, USA

GeniaLab BioTechnologie, Hamburger Str. 245, 38114 Braunschweig,
Germany (Fax: 49-531 23 21 0 22)

Hamilton, P.O. box 26, 7402 Bonaduz, Switzerland (Fax: 41-81 37 25 63)

Inotech AG, Kirchstrasse 1, 5605 Dottikon, Switzerland
(Fax: 41-56 624 29 88)

Institut für Mikrotechnik Mainz GmbH, Carl Zeiss-Str. 18-20, 55124 Mainz
Germany (Fax: 49-6131-990-205)

Kelco Int., Waterfield, KT20 5HQ Tadworth Surrey, U.K.
(Fax: 44-1-737 377 100)

Microdrop, Mühlenweg 143, 22844 Nordestedt, Germany
(Fax: 49-40-526 68 63)

Omya, 35, quai André Citroen, 75015 Paris 15, France
(Fax: 33-1-40 58 44 00)

Primex Ingredients, PO Box 114, 4262 Avaldsnes, Norway
(Fax: 47 52 85 70 77)

Pronova Biomedical, Gaustadalléen 21, 0371 Oslo, Norway
(Fax: 47-22 69 64 70)

Pronova Biopolymer, P.O. Box 494, 3002 Drammen, Norway
(Fax +47-32 20 35 10)

SKW Biosystems, B.P. 23, 85800 Isle sur Sorgue, France
(Fax: 33-4-90.21.31.86)

Sodeva, BP 299, 73375 Le Bourget du Lac cedex, France
(Fax: 33-4-79 26 12 65)

Wolff AG, Postbox 1515, 29655 Walsrode, Germany
(Fax: 49 5161 44 3376)

Measurement of Density, Particle Size and Shape of Support

ERIK VAN ZESSEN, JOHANNES TRAMPER, and ARJEN RINZEMA

Introduction

Density of gel beads

The density of gel beads is definied by the mass of a certain amount of gel beads divided by the corresponding volume. So to determine the density of gel beads, volume and mass have to be measured. Measurement of mass and volume of a certain amount of gel beads is not straightforward; gel beads consist in a large part of water, and consequently some water will adhere to the gel-bead surface. This amount of water adhering to the gel beads, is not part of the gel beads itself. This thin layer of water cannot be removed by simply sieving of these beads, and has to be accounted for in determining the density. This adhering amount of water can be measured by using a solution of a certain component that will not diffuse into the gel beads. The water adhering to the gel-bead surface will dilute this solution, and the amount of water follows from the dilution of this component.

Outline

The mass of the total amount of gel beads is measured and the amount of attached water is determined. Next, the real volume and mass of the gel beads is calculated and the density follows directly.

✉ Erik van Zessen, WageningenUniversity , Food and Bioprocess Engineering Group, Biotechnion PO Box 8129, Wageningen, 6700 EV, The Netherlands
(*phone* +31-0317-4-82240; *fax* +31-0317-4-82237;
e-mail Erik.vanZessen@algemeen.pk.wau.nl)
Johannes Tramper
Arjen Rinzema

Materials

Dextran Blue solution with a salt concentration equal to the salt concentration of the solution that is used to hold the gel beads.

Procedure

1. Weigh a calibrated flask, and fill this flask up to 40% of its volume with the Dextran Blue solution, and weigh again (= M_1).

2. Sieve the gel beads and fill the calibrated flask with these sieved gel beads up to almost its preset volume; weigh the flask (= M_2).

3. Fill the calibrated flask up to its preset volume, and weigh this flask (=M_3).

4. Shake this flask gently for at least 2 hours.

5. Determine the UV-absorption at 280 nm of the supernatant solution, and determine the UV-absorption of the Dextran Blue solution itself.

Calculation of the gel bead volume

Below is given the equation to calculate the true volume of the gel beads:

$$V_{\text{water added}} = \frac{M_1 - M_{\text{flask}} + M_3 - M_2}{\rho_{\text{water}}(T)}$$

$$V_{\text{watertootal}} = V_{\text{water added}} \cdot \frac{Abs_{\text{solution}}}{Abs_{\text{supernatant}}}$$

$$V_{\text{gel beads}} = V_{\text{flask}} - V_{\text{watertotaal}}$$

V_{flask}	volume of the flask (ml)
M_{flask}	mass of the flask (g)
M_1, M_2, M_3	mass as defined in the procedure part (g)
$\rho_{\text{water}}(T)$	density of the Dextran Blue solution at measured temperature (g/ml)
$Abs_{\text{supernatant}}$	UV absorption at 280 nm of the supernatant
Abs_{solution}	UV absorption at 280 nm of the Dextran Blue solution

The density of the gel beads is calculated with the next equation.

$$M_{water\ adhering} = (V_{watertotaal} - V_{wateradded}) \cdot \rho_{water\ (T)}$$

$$M_{gelbeads} = M_2 - M_1 - M_{water\ adhering}$$

$$\rho_{deeltjes} = \frac{M_{gelbeads}}{V_{gelbeads}}$$

Results

This paragraph gives an example of a single experiment as described above. It shows the influence of the volume of adhering water on the density of gel beads.

Volume of the flask	1000 ml
Massa of the flask (M_{flask})	278.34 g
M1	619.20 g
M2	1247.85 g
M3	1282.15 g

Temperature is 27 °C, which gives a density of the Dextran Blue solution of 0.9965 g/ml.
The UV-absorption of the solution itself: 0.984
The UV-absorption of the supernatant: 0.730
These measurements will give the following results:

$V_{water\ added}$	376.47 ml
$V_{water\ total}$	507.46 ml
$V_{gel\ beads}$	492.54 ml
$M_{water\ adhering}$	130.54 g
$M_{gel\ beads}$	498.12 g

So the density of the gel beads is 1.011 g/ml. It can be calculated that the water fraction adhering to the gel beads is 21%. Without accounting for this fraction of water the density becomes $(M_2 - M_1) / (V_{flask} - V_{water\ added}) = 628.65/623.53 = 1.008$ g/ml.

Shape of gel beads

The shape of a gel bead can be straightforwardly determined with any microscope. The resulting image can be digitalized and analysed with any appropriate software, e.g. Applied Imaging (Dukesway, Team Valley, Gateshead, Tyne & Wear, England).

Normally gel beads are close to a spherical shape. Their deviation from a spherical shape can be expressed as the elongation or circularity, two well-known shape factors. These shape factors are derived from different measurements done with the digitalized image.

Circularity is defined as : $\dfrac{4 \times \text{ bead area}}{\text{bead perimeter}^2}$, and the

Elongation is defined as : $\dfrac{\text{lenght bead}}{\text{breadth bead}}$.

The length of the bead is the distance between two points on the boundary of the gel bead which are furthest apart. The breadth of the bead is the distance between two points on the boundary of the gel bead which are furthest apart and normal to the length, see also Figure 1.

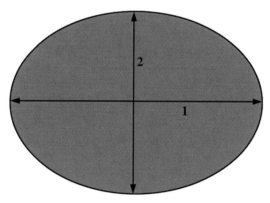

Fig. 1. Definition of object length (*1*), and object breadth (*2*)

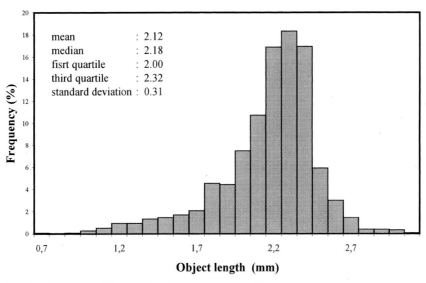

Fig. 2. Histogram of the length of κ-carrageenan gel beads

Diameter of gel beads

A rough estimate of the bead diameter can be obtained as follows: put a large number of beads along a ruler, and determine the total length as well as the number of beads. The mean diameter follows directly from dividing the total length by the number of beads.

As there are variations in gel bead diameter in any population of gel beads, one might characterise these variations. With a microscope a large number of gel beads are analysed with respect to the length and breadth. This can be done by digitalizing the image of the gel beads and analysing this image by appropriate software. Standard statistical techniques can be used to analyse the population, such as mean and median as a measure for the mean bead diameter, standard deviation and quartiles as a measure for the width of the population. As an example, Figure 2 shows a histogram of the length distribution of a population of κ-carrageenan gel beads, together with some characteristic numbers.

Acknowledgements. The Association of Biotechnology Centers in the Netherlands (ABON) is acknowledged for their financial support.

Mechanical Stability of the Support

Emily J.T.M. Leenen

Introduction

Industrial applications of immobilized biocatalysts have been gaining importance in the last decades. Some examples are given in the Chapter 21, Chapter 22 and Chapter 23. If these biocatalysts are used in reactors, such as aerated stirred tank reactors or air-driven reactors, the particles are continuously subjected to hydrodynamic shear stresses, motion and bursting of gas bubbles, and collisions against other particles and reactor parts. Severe abrasion of biocatalyst particles have been observed in these systems (Gjaltema et al., 1995, Hunik & Tramper, 1991). The effect of different kinds of mechanical stresses in reactors largely depends on the properties of the particle itself. Not only on the size, shape and density of the particle, but also ist roughness, hardness, elasticity, effect of fatigue and degree of homogeneity. These last properties determine how a biocatalyst particle is able to accommodate the stresses to which it is subjected in a reactor.

Several techniques can be used to measure rheological properties of support material. Mostly fracture properties (in compression) are measured, which gives information about fracture stress, strain at fracture, fracture energy and Young's modulus in compression. This however, does not give enough information about the resistance against abrasion in bioreactors. For this, oscillation experiments also have to be done to simulate the periodic application of stresses in a reactor and show the effects of fatigue or weariness of the material (related to the amount of abrasion).

Emily J.T.M. Leenen, National Institute for Public Health and the Environment, Microbiological Laboratory for Health Protection, P.O. Box 1, Bilthoven, 3720 BA, The Netherlands (*phone* +31-302473711; *fax* +31-30-2744434; *e-mail* Imke.Leenen@rivm.nl;frank-Imke@hetnet.nl)

Reology

A given material deforms if a force is applied. As reaction, the material will exert a certain back force. The extent to which the material deforms depends strongly on its intrinsic properties and on how (and for how long and at which rate) the forces are applied. An elastic material deforms immediately to a certain extent and returns to its original shape after the stress is removed. In viscous materials the deformation remains after the stress is removed. Natural gels and most of the synthetic polymers used for immobilization are visco-elastic materials, which means that the ratio of elastic to viscous properties depends on the time scale of the deformation. At short time scales their behavior is mainly elastic, whereas over long time scales the behavior contains a strong viscous component. For such materials stiffness and fracture stress depends on the rate at which the material is deformed (van Vliet & Peleg, 1991).

■ Materials

All materials used for immobilization can be used for the tests described below. For most tests it is better to use larger test pieces in the shape of cylinders, others can be done with gel beads (produced as described in (Chapter 2 and Chapter 3). Two methods to prepare cylindrical test pieces are:

Preparation of the support materials

- After preparation of the gel solution (depending on the support material, e.g. agar or carrageenan) this can be poured into a glass beaker or cylindrical cup. After gelation, gels can either be chopped into the correct shape or height or cut with a special borer to obtain the test pieces.

- After preparation of the gel solution this can be poured into a dialyses membrane. These membranes with solution can then be transferred to the appropriate solutions with counterions to obtain a homogenous gel (e.g. alginate). After gelation gels can either be chopped into the correct shape or height or cut with a special borer to obtain the test pieces.

Pieces with macroscopic structural defects should be rejected. For a rigorous analysis it is important to have homogeneous, reproducible gels that always exhibit the same properties under the same conditions.

▦ Procedure

Measurements of fracture properties

Force-compression curves should be determined at a constant temperature (e.g. $20 \pm 0.5°C$), using a tension-compression device (e.g. Zwick table model 142510) fitted with a 2000 or 50 N load-cell for testing cylindrical test pieces and gel beads, respectively. This apparatus consists of a fixed bottom plate and a bar containing the load cell, which can be moved at fixed speeds within the range of 0.1 - 500 mm.min^{-1}. The test pieces should be placed on the bottom plate and compressed with a given fixed speed. The force needed for deformation is recorded as a function of time until fracturing of the test piece. In this way a force-compression curve is obtained. Generally between 5 to 15 samples per batch need to be tested.

Creep measurements

To assess the visco-elastic behavior of support materials compression and recovery tests can be done with an apparatus described by Mulder (1946). Cylindrical test pieces are placed on a fixed plate and subjected to uniaxial compression due to a constant weight placed on top of the sample (e.g. $\pm$ 3000 N.m^{-2}). The deformation of the test pieces can be measured as a function of time by a displacement-transducer (e.g. Hewlett Packard). After some time the weight can be removed and the samples are allowed to recover, which can be measured again with the displacement-transducer.

Oscillation test

Oscillation tests are done to evaluate the effect of cyclic compressions on a support material. Test pieces are placed in a beaker (with an appropriate solution to avoid dehydration of hydrogels or leaking of counterions) mounted on the fixed plate of a tension-compression device (e.g. Overload Dynamics table model S100) fitted with a 2000 N load cell. Samples can be compressed at a fixed speed (e.g. 30 mm.min^{-1}) until a percentage (e.g. 25, 50 or 75%) of its deformation at fracture (determined from fracture tests), after which the moving bar returns to its original position. This movement can be repeated several times (e.g. 1000 times) and force-compression curves can be obtained.

Results

Fracture properties

Fracture of a material is basically the result of large deformations (strains) that ultimately lead to falling apart of the material into separate pieces. In viscoelastic polymers, fracture is accompanied by flow of (parts of) the material. This means that by applying a steadily increasing stress (force acting on a surface element divided by the area of that surface element) the material first starts to flow (locally) and fractures later. This behavior can be studied by constructing force-compression (stress-strain) curves. A typical example of such a curve is presented in Figure 1.

As the material is being compressed (strained) it exerts an increasingly high reaction force against the compressing agent. Beyond a certain (relative) deformation (or stress), the material can no longer resist and fractures. This point is given by the stress (τ_f in Figure A or G) and strain (ε_f in Figure B or H) at fracture respectively. The area below the curves gives the energy ($N/m^2 * m/m = Nm/m^2 = J/m^3$) needed to fracture the test piece (e.g.

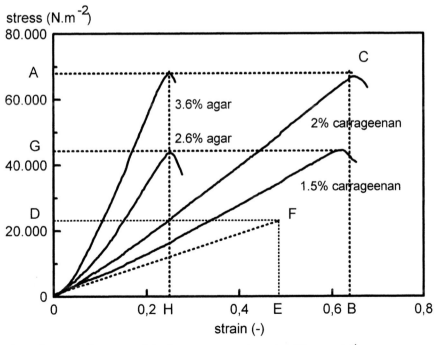

Fig. 1. Stress-strain curves of 4 gels at a compression rate of 3 mm.min^{-1}

0CB or 0GH in Figure) and includes the net energy required for fracture and the energy dissipated due to friction and flow of the material. A measure of the ratio between the stress exerted on a material and the relative deformation of the sample (in fact a measure of the reciprocal of the deformability) is the Young's modulus represented by the ration FE/DF in Figure 1. For most materials this ratio depends on the deformation itself. Only for (very) small deformations is this independent of it and representative for the undisturbed material.

Fracture stress, strain at fracture, fracture energy and Young's modulus in compression can thus be calculated on the basis of force-compression curves obtained.

Stresses ($N.m^{-2}$) can be calculated by dividing the force registered at every point by the corresponding bearing area, which is the area on which the force (moving plate) is acting. This area is changing constantly with the deformation. For cylindrical samples and assuming that the volume of the test pieces remains constant (which is reasonable within a timescale of a few minutes), the actual bearing area A_t, is given by: $A_t = A_0 * h_0/h_t$, where A_0 is the initial area (m^2), h_t is the height at time t (m) and h_0 is the initial sample height (m). For gel beads the stresses can be calculated considering the contact area as the area in a circle with a continuously increasing radius. This change in radius is given by $r^2 = r_0^2 - (r_0 - c)^2$ in which r_0 is the bead radius and c is the absolute value of vertical change in bead radius due to compression. For zero compression the bearing area is thus zero.

Strain (-) corresponds to the Hencky strain, defined as:

$$\varepsilon_H = ((h_0 + h)/h_0)$$

in which h_0 is the initial sample height and Δh the change in height. The Hencky strain is obtained by relating any strain increase (in an already strained sample) to the changed dimension of the sample. The point in the Figure at which the sample fractures (maximum of the stress-strain curve) is characterized by the (maximum) fracture stress (τ_f) on the Y-axis and the accompanying fracture strain (ε_f) on the X-axis.

The total **energy** required for (macroscopic) fracture is given by the area below the stress-strain curve until the fracture point. Mathematically (Van Vliet et al., 1992):

$$E_f = \int_{\varepsilon=0}^{\varepsilon=\varepsilon f} \tau \, \delta\varepsilon \quad (J.m^{-3})$$

The **Young's compression modulus** (E, $N.m^{-2}$) is the stress divided by the strain at very small deformations, which can be calculated from the slope of the curve at a relative strain of e.g. 5% (e.g. between a strain

Creep measurements

By placing a constant weight (e.g. 100 g) exerting a certain pressure, depending on the surface of the sample, on top of a cylindrical sample the deformation can be registered in time. After a certain period of time (e.g. 24 hours) the weight can be removed and the recovery of the sample measured too. Typical examples of such curves are represented in Figure 2.

Oscillation test

This test can be done to simulate, to a certain extent, the periodic application of stresses in a reactor. For this, gel samples are subjected to repeated (e.g. 1000) compressions below their fracture level. Figure 3 shows some typical examples of such curves.

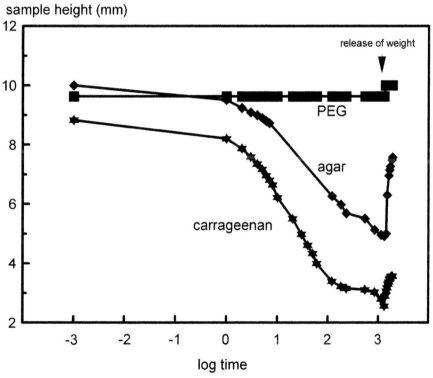

Fig. 2. Creep compression curves of 3 gels with a constant force of 3124 N.m^{-2}. After 33 hours the force was removed to determine the recovery

Generally the resistance to compression decreases with the number of oscillations.

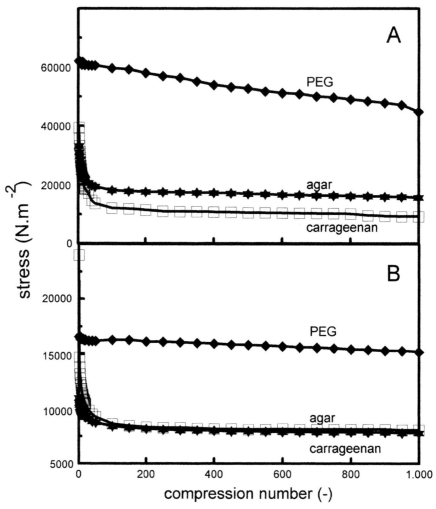

Fig. 3. Oscillation experiments of 3 gels. The evolution of the resistance of compression (stress) at a) 75% and b) 22.8% of their maximum compression. Compression rate of 3 mm.min⁻¹

References

Gjaltema A, Tijhuis L, Loosdrecht van MCM (1995) Detachment of biomass from suspended nongrowing spherical biofilms in airlift reactors, Biotechnol. Bioeng. 46: 258-269.

Hunik JH, Tramper J (1991) Abrasion of κ-carrageenan gel beads in bioreactors. Proc. Internat. Symp. Environmental Technology. Royal Society of Flemish Engineers, April 1991, Oostende, Belgium.

Leenen EJTM, Martins dos Santos VAP, Grolle KCF, Tramper J, Wijffels RH (1996) Characteristics of and selection criteria for support materials for cell immobilization in wastewater treatment. Water Res. 30: 2895-2996.

Martins dos Santos VAP, Leenen EJTM, Ripoll MM, Van der Sluis C, Van Vliet T, Tramper J, Wijffels RH (1997) Relevance of rheological properties of gel beads for their mechanical stability in bioreactors.Biotechnol. Bioeng. 56-5: 517-529.

Mulder H (1946) Het bepalen van reologische eigenschappen van kaas. Verslag Landbouwkundig Onderzoek (in Dutch), 51.

Van Vliet T, Peleg M (1991) Effects of sample size and preparation, In: Rheological and fracture properties of cheese. Bul. Int. Dairy Fed. 268.

Van Vliet T, Luyten H, Walstra P (1992) Time-dependent behavior of food. In: Food colloids and polymers: stability and mechanical properties. Dickenson & Walstra (eds.) Royal Society of Chemistry, Cambridge, UK.

Diffusion Coefficients of Metabolites

EVELIEN E. BEULING

Introduction

The conversion rate of most immobilized cells depends not only on the availability of substrates and microbial kinetics, the diffusion rate of metabolites through the immobilized cell system is of major importance as well. Therefore, characterization of the internal mass transfer properties may be considered essential for the modelling, design and scale-up of processes utilizing immobilized cells.

Several experimental approaches may be used for the determination of diffusion coefficients of immobilized cell supports. If the conversion rate is known at different substrate concentrations the diffusion coefficient can be calculated using a diffusion-reaction model. For that purpose, detailed knowledge of the microbial kinetics and cell loading is required, while internal biofilm conditions, like the occurrence of pH-gradients, have to be quantified (De Beer et al., 1992). Due to the inhomogeneous structure of biofilms and the poorly defined kinetics of immobilized micro-organisms, these results are not reliable and show large variations (Fan et al., 1990). Therefore these methods are not treated here.

More reliable diffusion coefficients may be obtained if non-reacting compounds are used. For that purpose, the diffusion of a non-reacting compound as for example nitrous oxide or a lithium salt can be monitored (Westrin & Axelsson, 1991; Libicki et al., 1988). The resemblance of these compounds to microbial metabolites is rather questionable, however. The use of inactivated biofilms enables diffusion experiments with actual metabolites. Inactivation can be achieved by incubation with a toxic compound (Chresand et al., 1988), or a heat shock (Tatevossian, 1985). These

Evelien E. Beuling, MARS B.V. R&D, Postbus 31, 5460, BB Veghel, The Netherlands (*phone* +31-413-383668; *fax* +31-413-351670; *e-mail* Evelien.Beuling@eu.effem.com.)

treatments may affect the mass transfer properties of biofilms (Lens et al., 1993) and results of such experiments have to be treated with care.

An ideal method for the measurement of diffusion coefficients in immobilized cell systems does not exist and depending on the shape of the immobilized cell support and the required precision, the best method has to be chosen. In this chapter two different methods will be presented:

- the pseudo steady-state diffusion cell,

- the step-response method utilizing micro-electrodes.

Prior to the detailed description of these two methods a short theoretical exposé will be given on diffusion in general.

Other interesting methods are available but mostly only the shape of the cell support differs; e.g. beads in a finite liquid volume or diffusion into an infinite slab (Westrin, 1991). All these methods, however, calculate the diffusion from a step-response and their theoretical background is similar to the step-response method treated here. Some other interesting new techniques have been reported like pulsed field gradient NMR (Beuling et al.,1998) and holographic laser interferometry (suited for pure gel materials only) for example (Westrin, 1991). Although application of these methods may be advantageous for immobilized cell systems, the experimental set-up is costly and specialized knowledge is required. Therefore, these methods are not treated here.

Diffusion

Diffusion is the process by which concentration gradients in a solution spontaneously decrease due to the Brownian motion of the individual molecules. An immobilized cell system may be considered as a heterogeneous system containing a continuous aqueous phase, and a dispersed phase, i.e. the micro-organisms and the polymer network. The fraction of the dispersed phase ϕ_d is generally assumed inaccessible to the solvent and the diffusant. Analogously to porous chemical catalyst particles, mass transfer in this system is characterized with an effective diffusion coefficient D_{eff} that is defined by:

$$J = -D_{eff} \cdot \frac{dC_c}{dx} \tag{1}$$

where J is the flux through the biofilm, and C_c the solute concentration in the liquid phase. Compared to the diffusion coefficient in a single phase liquid system, the value of D_{eff} is reduced by two different mechanisms:

i. the cross-sectional area available for diffusion is reduced to a fraction $(1-\phi_d)$ due to the presence of the dispersed phase; this is usually referred to as exclusion.

ii. the pathlength of diffusing molecules, referred to as the tortuosity τ, is increased due to the obstruction imposed by the inaccessible dispersed phase.

Theoretically, the effective diffusion coefficient in a dispersion can be related to the diffusion coefficient in the liquid phase D_c according to:

$$D_{eff} = \frac{(1 - \phi_d)}{\tau^2} \cdot D_c \tag{2}$$

The square of the tortuosity is required because dispersed particles not only increase the pathlength of the diffusing molecules, but also decrease the steepness of the concentration gradient experienced along this path (Epstein, 1989). Eq. 2 holds as long as there are no specific interactions between the diffusing solute and the dispersed phase, but has a limited value since both ϕ_d and τ of a biofilm are usually unknown. Hence, the determination of D_{eff} still has to be done experimentally.

Classical diffusion cell

In a classical diffusion, a (nearly) constant concentration difference over a membrane results in a steady-state flux which is monitored and used to calculate the effective diffusion coefficient of the membrane. At the beginning of this century, the diffusion cell equipped with a membrane of porous glass was frequently used for the study of diffusion phenomena. Later on, this method was replaced by more sophisticated techniques, like Rayleigh interferometry, and nowadays the diffusion cell is only utilized for the study of diffusion in materials like gels and plastics.

The diffusion cell consists of two compartments filled with well-mixed solutions (see Fig. 1). These two compartments are separated by a membrane. A diffusion experiment is started, by adding the tracer compound to one of the compartments. Hereafter, the solute penetrates from one side into the membrane. After a certain interval of time t_0; the so-called time-lag, the membrane is loaded with solute such that a linear concentration gradient within the membrane is established.

If the volume of the membrane is much smaller than the volume of both compartments, the adaptation time of the concentration in the membrane is much shorter than the time needed to induce a change of the concen-

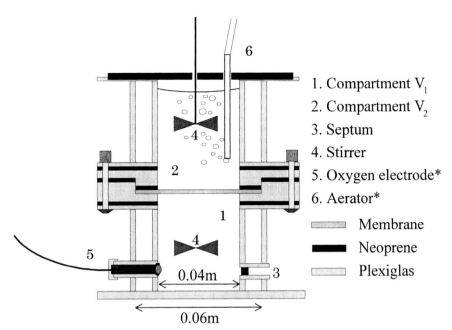

1. Compartment V_1

2. Compartment V_2

3. Septum

4. Stirrer

5. Oxygen electrode*

6. Aerator*

◤◣ Membrane

■■ Neoprene

▭ Plexiglas

Fig. 1. The experimental set-up of the diffusion cell. The two parts denoted with an asterisk are specifically required during experiments with oxygen

tration in one of the compartments. In that case, the flux into the membrane may be assumed equal to the flux out of the membrane and diffusion coefficient in the membrane may be derived from a relatively small concentration change in one compartment during a limited time interval. This procedure is referred to as pseudo steady-state and may be applied if the ratios between the diaphragm volume and the compartment volumes are less than 0.1 (Barnes, 1934).

If (pseudo) steady-state is attained, the amount of solute N diffusing through the membrane with thickness L and effective area A is determined by the concentration difference over the membrane according to:

$$N = \frac{D_{eff} \cdot A}{L} \cdot (C_1 - C_2) = N \tag{3}$$

with C_1 and C_2 as the concentrations in both compartments, respectively. The diffusing solute induces a concentration change in both compartments:

$$-V_1 \cdot \frac{dC_1}{dt} = +V_2 \cdot \frac{dC_2}{dt} = N \tag{4}$$

in which V_1 and V_2 denote the volumes of both compartments. The time-dependent concentration difference between both compartments may be described by substitution of Eq. 3 into Eq. 4, i.e.:

$$\frac{d(C_1 - C_2)}{dt} = \frac{D_{eff} \cdot A}{L} \cdot (C_1 - C_2) \cdot (\frac{1}{V_1} + \frac{1}{V_2}) \tag{5}$$

This differential equation can be solved with the boundary conditions:

$$t = t_0 \quad C_1 = C_{1,0} \tag{6a}$$

$$C_2 = C_{2,0} \tag{6b}$$

and yields:

$$\ln(\frac{C_1 - C_2}{C_{1,0} - C_{2,0}}) = -\frac{D_{eff} \cdot A}{L} \cdot (\frac{1}{V_1} + \frac{1}{V_2}) \cdot (t - t_0) \tag{7}$$

Plotting the experimental value of the left-hand side of this equation as a function of time, the effective diffusion coefficient D_{eff} may be obtained from the slope of the straight line (see Fig. 2).

Subprotocol 1
Preparation of the Membrane

The membrane has to be sufficiently rigid to enable assembly in the diffusion cell. In our laboratory, all experiments were performed with agar gels. In literature, however, it is reported that other gel types, like alginate and carrageenan, are applicable as well (Westrin & Axelsson, 1988).

Materials

- Buffer (50 mM K_2HPO_4: pH 7.0)
- Agar (Merck, Germany)
- Cell culture suspension (re-suspended in buffer)
- Horizontal circular metal mould ($6.0 \cdot 10^{-2}$ m diameter)
- Petri-dish of glass
- Ice
- Thermostatic hot plate

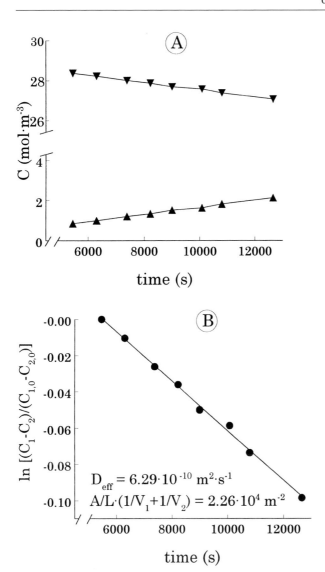

Fig. 2. Typical results of a glucose experiment in a diffusion cell. **A** glucose concentrations in compartment V_1 () and V_2 (), respectively. **B** logarithmic concentration difference between both compartments as a function of time

▓▓ Procedure

1. Mix the agar powder with buffer liquid and heat it in an autoclave at 120°C for 2 minutes. The maximum amount of agar that can be dissolved is ca. 8 % w/w.

2. Cool the liquid agar down to 50°C and mix it with the bacterial suspension of the same temperature.

3. Warm the mould on the thermostatic hot plate upto 50°C.

4. Pour the liquid agar suspension in the mould.

5. Cover the mould by the petri-dish.

6. Cool the mould with ice.

7. Take the solidified membrane carefully from the mould.

8. Inactivate the immobilized bacteria by incubating the membrane in buffer liquid containing 0.2% w/w $HgCl_2$, for 15 h.

9. Wash the membrane twice by an incubation in buffer liquid without $HgCl_2$ for two hours.

10. The membrane has to be stored in buffer liquid.

▓▓ Comments

For a reliable calculation of the diffusion coefficient, the thickness L of the gel membrane has to be exactly known and may be determined with an electrometer that monitors the resistance between a copper electrode (tip diameter 2 mm) fixed in a micro-manipulator (Uhl, Switzerland) and a flat metal plate. The resistance between the electrode-tip and the flat metal plate sharply decreases when the electrode tip contacts the metal surface or a gel membrane positioned on this plate. The difference of two such measurements equals to the thickness of the membrane. The average of six measurements at different positions of one membrane was taken as the thickness.

The thickness can be measured as well by microscopical observation of membrane cross-sections or measurements by an ultrasonic meter with a piezoelectric probe (Axelsson & Persson, 1988).

Subprotocol 2
Diffusion Experiments

▓ ▓ Materials

The experimental apparatus applied for our research is constructed from **Diffusion cell**
plexiglas and assembled from four separate parts as shown in Fig. 1. The
three upper parts contain 6 holes in the outer edge and they fit upon the
lower part which contains 6 bolts at the same place. To prevent leakage,
packing material with a low oxygen solubility (Neoprene) was laid in between.

The two cylindrically shaped compartments V_1 and V_2 contain $60 \cdot 10^{-6}$
m^3 of liquid each and have an internal diameter of $4.0 \cdot 10^{-2}$ m. These are
separated by a circular gel membrane with a thickness of $2.0 \cdot 10^{-3}$ m and
diameter of $6.0 \cdot 10^{-2}$ m. The $1.0 \cdot 10^{-2}$ m edge of the membrane is clamped
between two membrane holders of the diffusion cell.

Both compartments are surrounded by a water jacket that is connected
with a water bath of $30.0 \pm 0.2 °C$. The water jacket was only used to keep the
liquid at a constant temperature during the experiment, since the heat con-
duction of Plexiglas is very poor. For that reason the liquid for both compartments was brought to the right temperature before assembly of the
apparatus. During an experiment the temperature in the upper compartment was regularly checked.

The lower compartment contains a magnetic stirrer ($2.5 \cdot 10^{-2}$ m). The
rubber septum (diameter $1.0 \cdot 10^{-2}$ m) was used for the injection of tracer
solute and adjustment of the liquid volume. The other gap was used to
insert an oxygen electrode or a conductivity cell.

The upper compartment is equipped with an electric stirrer. The speed
of the stirrers in both compartments was set high enough (500 rpm) to
ensure that mass transfer resistance in the liquid phase can be neglected.
In order to verify whether the mixing intensity in both compartments was
sufficient, experiments with reduced stirrer speeds were performed. A reduction to 50% effected no decrease of the measured diffusion coefficient
indicating that the assumption regarding the mass transfer resistance in
the liquid phase was justified.

– Gel membrane containing inactivated immobilized cells   **Reagents,**
                                                                    equipment
– Buffer (50 mM K_2HPO_4: pH 7.0)

- 1.5 M glucose solution (density: 1097 kg · m⁻³)

- 5-ml-syringes

- Enzyme kit for the determination of the glucose concentration (Sigma diagnostics, St. Louis, USA)

- Spectrophotometer

Procedure

The diffusion cell can be used for the determination of diffusion coefficients of many tracer compounds. There are two basic requirements. The concentration of the diffusing solute has to be measured with a sufficient accuracy and without consuming large sample volumes. The volume of one sample must be 0.5% or less of the volume of one compartment or should be monitored in-line (as for example oxygen or sodium chloride with an electrode or conductivity cells, respectively). In literature, measurements with other sugars, organic acids, ethanol and ureum have been described. In this chapter an experiment with glucose will be presented and experiments with oxygen and sodium chloride are described briefly.

Diffusion experiment

1. Degass the buffer liquid for the lower compartment by a suction pump. In this way formation of gas bubbles can be avoided.

2. Fill the lower compartment with buffer liquid of about 30°C.

3. Place the membrane on the liquid surface.

4. Place the upper membrane holder and fasten the 6 screws equally without damaging the clamped edges of the membrane.

5. Adjust the volume in the lower compartment through the septum with a syringe filled with buffer liquid, such that the surface of the gel is flat.

6. Fill the upper compartment with buffer solution of 30°C.

7. Withdraw $1.0 \cdot 10^{-6}$ m³ buffer liquid with a 5 ml syringe from the lower compartment V_1 and start the experiment by injecting $1.0 \cdot 10^{-6}$ m³ of a 1.5 M glucose solution (density: 1097 kg · m⁻³).

8. Determine the amount of injected solution exactly by weighing the syringe containing the glucose solution before and after the injection.

9. Wait for ca. 100 min. ($\approx 0.85 \cdot L^2/D_{eff}$) to establish pseudo steady-state.

10. Collect 8 samples of $0.25 \cdot 10^{-6}$ m^3 from the upper compartment V_2 in the next two hours.

11. Immediately after finishing the experiment, tap a sample from the lower compartment V_1 ($C_{1,end}$).

12. Determine the amount of liquid in both compartments by weighing.

13. Check the membrane visually, if any cracks or fissures can be observed the experiment is not reliable.

14. Measure the glucose concentration C in the samples spectrophotometrically with the enzyme kit.

15. Calculate the diffusion coefficient from the time-dependent concentration difference (Fig. 2). The concentration in the lower compartment is estimated from the experimentally determined $C_{1,end}$.

Results

Typical results of a glucose experiment are shown in Figure 2.

Comments

Experiments with oxygen were performed in the same apparatus. For that purpose, an oxygen probe (Yellow Springs Instruments, Ohio, USA) was fixed in the lower compartment and both compartments were filled with nitrogen-saturated buffer liquid. An aerator flushed the liquid in the upper compartment with nitrogen. The experiment was started by replacing the nitrogen supply by air. Thus, the oxygen concentration in the upper compartment changed within 600 s to $C_{2,0}$, i.e. $2.44 \cdot 10^{-1}$ mM (Perry & Green, 1988). After a lag-time of ca. 30 min ($\approx 0.85 \cdot L^2/D_{eff}$), the oxygen concentration in the lower compartment was monitored during the next 2h period. The effective diffusion coefficient was calculated with:

Other tracer compounds

$$\ln\left(\frac{C_1 - C_{2,0}}{C_{1,0} - C_{2,0}}\right) = -\frac{D_{eff} \cdot A}{L} \cdot \frac{1}{V_1} \cdot (t - t_0) \tag{8}$$

The diffusion rate of sodium chloride was measured by positioning two conductivity cells in the upper and lower compartment, respectively. These experiments were performed with pure water instead of buffer liquid to enhance the accuracy of the conductivity measurements. For that reason, these experiments were performed with plain ager gels only.

The edge effect   The clamped edges of the membrane are not directly connected to the liquid compartments and during the first part of an experiment, the loading of the clamped edges causes a decrease of the flux through the membrane in comparison with a two-dimensional situation without edges. Therefore, a relatively long lag-time is required. The length of this period was experimentally determined from test-experiments with sodium chloride which could be detected continuously in both compartments with conductivity cells. It was shown that the flux in and out of the membrane approximated each other within 2% after about $0.85 \cdot L^2/D_{eff}$. Numerical calculation showed that in the two-dimensional case without edges $0.5 \cdot L^2/D_{eff}$ should be sufficient to attain pseudo steady-state.

Once steady-state in the entire membrane is established, an extra flux through the edges will occur. Thus, in comparison with the two-dimensional situation the observed flux will be larger. Barrer et al. (1962) analytically calculated that the steady-state flux increases by about 2% due to the presence of clamped edges with the dimensions applied here. However, steady-state is only approached very slowly and according to Westrin (1991) the increase of the flux is opposed by the loading of these edges during the period between $0.8 \cdot L^2/D_{eff}$. and $1.5 \cdot L^2/D_{eff}$. Since our samples were collected during this period, these side effects were not accounted for.

Subprotocol 3
Step-Response Method Utilizing Micro-Electrodes

The diffusion coefficient can be calculated from the transient response on a concentration change of a non-metabolized compound that is monitored with a micro-electrode positioned in the center of a spherical biofilm. With this method the diffusion coefficient is obtained from the non-steady state flux into the immobilized cell system and does not yield an effective diffusion coefficient as given in Eq. 1. It is essential to distinguish this and it can be explained by a microscopic balance over the heterogeneous material with C_c concentration in the continuous liquid phase:

$$(1 - \phi_d) \cdot \frac{\partial C_c}{\partial t} = \frac{D_{eff}}{r^2} \cdot \frac{\partial}{\partial r}\left(r^2 \frac{\partial C_c}{\partial r}\right) \tag{9}$$

in which ϕ_d denotes the volume fraction of the dispersed phase, i.e. the cells and the polymer network. The fluxes into and out of the slice are determined by the effective diffusion coefficient, but the resulting concentration change in the continuous phase depends on the fraction of the liquid accessible for the diffusant, i.e. the porosity $(1-\phi_d)$. As a consequence, the concentration develops according to:

$$\frac{\partial C_c}{\partial t} = \frac{D_{eff}}{(1-\phi_d)} \cdot \frac{\partial}{\partial r} \cdot \left(r^2 \frac{\partial C_c}{\partial r} \right) \tag{10}$$

These equations are based on the assumption that no interactions exist between the diffusing solute and disperse phase. In case the diffusing solute adsorbs to or dissolves in the dispersed phase, more tracer compound has to diffuse into the gel bead before equilibrium is reached and a delay of the response in the centre of the gel bead is effected. This has to be accounted for in the mass-balance:

$$(1-\phi_d) \cdot \frac{\partial C_c}{\partial t} + \phi_d \cdot \frac{\partial C_d}{\partial t} = \frac{D_{eff}}{r^2} \cdot \frac{\partial}{\partial r} \cdot \left(r^2 \frac{\partial C_c}{\partial r} \right) \tag{11}$$

in which C_d represents the concentration in the dispersed phase. The concentration in the dispersed phase may be assumed to be continuously in equilibrium with the concentration in the surroundings, then C_d equals $m \cdot C_c$. with m as the partition coefficient between both phases. So the apparent diffusion coefficient D_{app} obtained is related to the effective diffusion coefficient by:

$$D_{app} = \frac{D_{eff}}{1 + \phi_d \cdot (m-1)} \tag{12}$$

Hence, an accurate estimation of D_{eff} from transient experiments is only possible if the porosity $(1-\phi_d)$ of the biofilm is known and partition phenomena are properly quantified. This is not always recognized in the literature and, consequently, diffusion coefficients obtained from transient measurements are often erroneously presented as effective diffusion coefficients (Westrin & Axelsson, 1991; Fu et al., 1994). Furthermore, it is not possible to deduce the diffusive permeability of the dispersed phase with transient experiments. To that end, steady-state experiments have to be performed (Beuling et al., 1996).

To describe the concentration curve in a spherical immobilized cell system with radius R at position r, the appropriate initial and boundary conditions have to be used:

$$0 \leq r \leq R \qquad C = C_0 \qquad t = 0 \tag{13a}$$

$$r = R \qquad C = C_1 \qquad t \geq 0 \qquad \qquad (13b)$$

$$r = 0 \qquad dC/dr = 0 \qquad \qquad (13c)$$

and the general solution of this set of equations yields (Crank, 1975):

$$\frac{C - C_0}{C_1 - C_0} = 1 + \frac{2R}{\pi r} \sum_{n=1}^{\infty} \frac{(-1)^2}{n} \cdot \sin\left(\frac{n\pi r}{R}\right) \cdot \exp\left(-n^2\pi^2 \frac{D_{app}t}{R^2}\right) \qquad (14)$$

The concentration response in the center of the bead is given by the limit of $r \to 0$:

$$\frac{C - C_0}{C_1 - C_0} = 1 + 2 \sum_{n=1}^{\infty} (-1)^n \cdot \exp\left(-n^2\pi^2 \frac{D_{app}t}{R^2}\right) \qquad (15)$$

This micro-electrode method is suited as well to measure diffusion coefficients of metabolites in active biofilms. For that purpose, the substrate concentrations during the entire experiment have to be sufficiently high compared to the Monod constants of the immobilized micro-organisms. In that case, zero-order activity will prevail in the entire biofilm and external concentration changes are transmitted unattenuated into the biofilm (Beuling et al., 2000). The same equations as with inactivated biofilms are used to describe the concentration change in a biofilm with a constant activity.

Materials

Preparation of gel beads containing immobilized cells

Gel beads   The gel material has to be sufficiently soft to enable penetration with a micro-electrode. Gel types like carrageenan and alginate for example are too tough although measurements in such gels containing a higher cell load have been possible. In this contribution, only measurements with agar are described.

An agar solution of 50°C was produced as described with the membrane preparation for the diffusion cell. The liquid agar was mixed with a diluted bacterial suspension and trickled through a Pasteur pipette in highly fluid paraffin oil of 4°C. The drops solidified in the cold paraffin and spherical gel beads with a diameter of ca. 4 mm were obtained. The sphericity of the gel beads never differed more than 1% from unity. These beads were rinsed extensively with a buffer solution and treated with $HgCl_2$ in case inactivation of the immobilized cells (see membrane preparation) is required. They were stored at 6°C in buffer liquid.

The glucose micro-electrode consists of a glass coated platinum wire (tip diameter 10 μm). The tip of the micro-electrode is covered with an enzyme layer containing glucose oxidase that converts glucose and oxygen into hydrogen peroxide and gluconate. If the platinum surface of the micro-electrode is positively polarized versus a Ag/AgCl-reference macro-electrode (0.7 V), the hydrogen peroxide is reduced at the platinum surface, and results in an electric current proportional to the glucose concentration upto 10 mM. The typical response at 30°C is 50 pA/mM, and the response time (τ_{90}) is less than 5 s. Sensor preparation is described elsewhere (Cronenberg & Van den Heuvel, 1991; Cronenberg, 1994).

Micro-electrodes

The combined oxygen electrode consists of a glass capillary (tip diameter 10 μm) containing a glass coated platinum wire and a Ag/AgCl-reference electrode. The capillary is filled with a 4 M KCl solution and the tip is sealed with silicone rubber. The electrode has been described by Revsbech and Ward (1983). The sensitivity at 30°C is 2 nA/mM and the response time (τ_{90}) is less than 2 s.

The experimental set-up for the micro-electrode measurement

The experimental set-up is shown in Fig. 3 and consists of a rectangular Plexiglas vessel with inner dimensions h×w×d of 3.7×10.0×3.5 cm³ filled with 100 · 10⁻⁶ m³ buffer liquid. The entire set-up is placed in a Faraday cage to reduce electromagnetic interference from the surroundings. The gel bead is fixed in the center of the vessel with entomological specimen needles on a holder of rubber. The temperature is kept at 30.0 ± 0.1 °C by a thermostat consisting of a resistance in a tube of glass provided with voltage by a rechargeable battery positioned outside the Faraday cage. Mixing is accomplished by bubbling nitrogen, air or oxygen through the solution ($8.3 \cdot 10^{-6}$ m³ · s⁻¹). The mixing intensity has to be sufficient to ensure that mass transfer resistance in the liquid phase can be neglected. In order to verify whether the applied gas flux was high enough, experiments with reduced gas fluxes were performed. A reduction to 30% effected no decrease of the measured diffusion coefficient indicating that the assumption regarding the mass transfer resistance in the liquid phase was justified.

Experimental set-up

The micro-electrode is positioned in one direction (arrow in Fig. 3) with an accuracy of 2 μm by a motor-driven micro-manipulator (Uhl, Switzer-

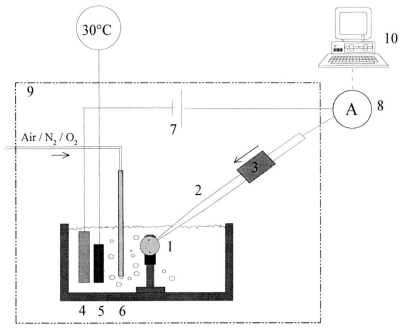

Fig. 3. Experimental set-up for a micro-electrode measurement. *1* Agar bead; *2* Micro-electrode; *3* Micro-manipulator; *4* Reference electrode; *5* Thermostat; *6* Aerator; *7* Voltage source; *8* Electrometer; *9* Faraday cage; *10* Computer

land). The holder of the micro-electrode is positioned on a hand-geared transmission and can be moved at the three retangular axes.

The current is measured with an electrometer (Keithley model 617, Cleveland, USA) connected to an IAAA-interface of a computer. Data sampling is done as follows: seven successive values of the current are collected in 7 s, after rejection of the two lowest and highest values of this data-set the three remaining values are averaged. In this way, possible spikes in the signal are discarded.

Further reagents and equipment

- Glucose micro-electrode

- 1 M glucose solution

- Buffer (50 mM K_2HPO_4: pH 7.0)

- Horizontal microscope provided with a graduated eye piece allowing an accuracy of 30 μm

▉ ▉ Procedure

Performance of a diffusion experiment

The range of solutes that can be measured with micro-electrodes is limited; so far only the diffusion rate of glucose, oxygen and hydrogen peroxide have been measured. If micro-electrodes for other solutes are available, however, determination of the diffusion coefficient of these compounds should be possible as well.

1. Fill the vessel with 100 ml buffer liquid of ca. 30°C.

2. Fix the gelbead with the entomological specimen needles on the rubber and place the holder in the vessel.

3. Measure the diameter of the fixed gel bead in-situ with the microscope. Take the average of three different measurements as the bead diameter.

4. Fix the micro-electrode in the motor driven micro-manipulator.

5. Position the micro electrode perpendicular on the surface of the gel-bead with the aid of the graduation of the microscope and the depth of field.

6. Position the tip of the electrode exactly on the bead surface and set the manipulator on zero.

7. Penetrate the gel bead with the micro-electrode until the tip is positioned in its center.

8. Wait for ca. 30 min. to stabilize the signal and sample the zero signal, i.e. C_0.

9. Add 1 ml of 1M glucose and start sampling.

10. Monitor the concentration development inside the bead. A period of ca. $0.9 \cdot R^2/D$ should be sufficient to get the entire diffusion curve and the end concentration C_1. In case the electrode signal shows some slow drift due to electrode deactivation, a longer period should be sampled. Typical concentration curves are given in Fig. 4.

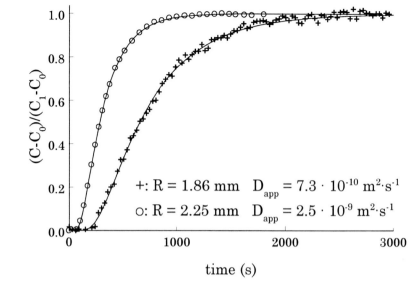

Fig. 4. The response measured with a micro-electrode in the centre of 1.5 % w/w agar beads: glucose (+) and oxygen (○). Solid lines: theoretical relation according to Eq. 15, using the appropriate bead radius and best diffusion coefficient

Calculation of the diffusion coefficient

The signal may be corrected for a slow drift due to electrode deactivation. If the signal reduction exceeds 4% per hour, electrodes are not considered suitable. The diffusion coefficient is obtained by fitting the recorded concentration curve with Eq. 15. The fit with the least sum of squares has to be taken.

Experiments with oxygen

Similar experiments with oxygen can be performed as well. The well-defined concentration change in the bulk liquid, however, is more difficult to establish. If only the composition of the gas phase that is flushed through is changed at t=0, the new equilibrium concentration in the bulk liquid is not established instantaneously due to the finite mass transfer rate from the gas to the liquid phase. This mass transfer rate can be characterized as a first-order process and, consequently, the time-dependent concentration in the bulk liquid is described by:

$$\frac{C_b - C_{b,0}}{C_{b,1} - C_{b,0}} = 1 - e^{-t/\beta} \tag{16}$$

with β representing the time constant of the saturation process. The concentration change may be monitored with an electrode positioned in the bulk liquid. With our experimental set-up we found a concentration curve that was satisfactorily described by a time constant β of 43.5 s (Beuling et al., 1999). Neglectance of this delay under the applied experimental conditions gives an experimental diffusion coefficient that underestimates the real diffusion coefficient with approximately 10%.

A correct fit equation is obtained if Eq. 16 is used as the boundary condition for r=R and $t \geq 0$. This yields the following solution at r = 0 (Crank, 1975):

$$\frac{C - C_{(0,0)}}{C_{(0,\infty)} - C_{(0,0)}} = 1 \frac{\sqrt{(R^2\beta/D_{app})} \cdot \exp(-t/\beta)}{\sqrt{\sin(R^2/\beta D_{app})} - ...}$$

$$... + \frac{2R^2}{\beta D_{app}} \sum_{n=1}^{\infty} (-1)^n \frac{\exp(-D_{app}n^2\pi^2t/R^2)}{n^2\pi^2 - R^2/\beta D_{app}} \tag{17}$$

A stepwise change ($\tau_{90} \leq 5$ s) of the oxygen concentration from anoxic to air-saturated conditions is effected by stripping the contents of the measuring vessel with N_2 prior to the addition of oxygen saturated buffer solution up to air saturation. Simultaneously, the supply of nitrogen is stopped, and exchanged for air.

Comments

The precision of the diffusion coefficient obtained with a micro-electrode is mainly determined by the accuracy of the measured bead radius R. This value has such a large influence since its square is used in Eq. 15. The diameter of this bead is determined using a microscope with an accuracy of 30 μm, so the measured value of a typical gel bead with a radius of 1.75 mm differed at most 15 μm, i.e. 0.8%. This corresponds with an error of 1.6% in the calculated diffusion coefficient.

Other important parameters that influence the precision of the measurement are the position r of the electrode and its response time. The electrode is positioned perpendicular to the bead surface with the aid of a microscope. However, the depth of field of the microscope is limited, introducing an error of about 200 μm in the final position of the electrode-tip. In a gel bead with a radius of 1.75 mm this erroneous position results in an overestimation of 0.8% of the real diffusion coefficient. Retardation of the signal (Δt), due to a slow electrode response, is at most $10^{-3} R^2/D$, which introduces a 0.3 % underestimation of the real diffusion coefficient.

The variance of diffusion coefficients determined with micro-electrodes is small compared to another generally used transient technique, viz. the stirred bath technique (Westrin, 1991). According to this method, the concentration change in the liquid containing submerged gel beads is measured after a concentration step. Such an experiment requires a high volume fraction of mono-disperse gel beads (Westrin & Zacchi, 1991), otherwise the concentration differences are too small. However, this hinders effective mixing (Ruggeri et al., 1991). These disadvantages do not apply for the micro-electrode method. The diffusional behavior of a single gel bead is measured. Hence, a batch of gel beads with a well-described size distribution is not required, and the mixing intensity of the bulk liquid can be chosen high enough to assure a negligible external mass transfer resistance. Furthermore, the monitored concentration change is directly proportional to the concentration step applied in the bulk liquid.

Acknowledgements. The material presented in this contribution was gained during a PhD-study at the Department of Chemical Technology, University of Amsterdam, Holland. The supervisors Prof. ir. S.P.P. Ottengraf and Dr. J.C. Van den Heuvel are gratefully acknowledged for their constructive comments and interesting discussions.

References

Axelsson A, Persson BA (1988) Determination of effective diffusion coefficients in calcium alginate gel plates with varying yeast cell content. Appl. Biochem. Biotechnol. 18:231-250

Barnes C (1934) Diffusion through a membrane. Physics 5:4-8

Barrer RM, Barrie JA, Rogers MG (1962) Permeation through a membrane with mixed boundary conditions. Trans. Far. Soc. 54:2473-2483

Beuling EE, Van den Heuvel JC, Ottengraf SPP (1996) Determination of biofilm diffusion coefficients using micro-electrodes. In: Wijfels et al.. (Eds.), Immobilized cells: basics and applications. Prog. Biotechnol. 11: 31-38

Beuling EE (1998) Mass transfer properties of biofilms. PhD-thesis, University of Amsterdam (ISBN 90-5651-058-4)

Beuling EE, Van Dusschoten D, Lens P, Van den Heuvel JC, Van As H, Ottengraf SPP (1998) Characterization of the diffusive properties of biofilms with pulsed field gradient-nuclear magnetic resonance. Biotechnol. Bioeng. 60:283-291

Beuling EE, Van den Heuvel JC, Ottengraf SPP (2000) Diffusion coefficients of metabolites in active biofilms. Biotechnol. Bioeng. 67:53-60

Crank J (1975) The mathematics of diffusion (2nd ed). Clarendon Press, Oxford Cronenberg CCH, Van den Heuvel JC (1991) Determination of glucose diffusion coefficients in biofilms with micro-electrodes. Biosens. Bioelectron. 6:255-262

De Beer D, Huisman JW, Van den Heuvel JC, Ottengraf SPP (1992) The effect of pH profiles in methanogenic aggregates on the kinetics of acetate conversion. Wat. Res. 26: 1329-1336

Epstein N (1989) On tortuosity and the tortuosity factor in flow and diffusion through porous media. Chem. Eng. Sci. 44:777-779

Fan L-S, Leyva-Ramos R, Wisecarver KS, Zehner BJ (1990) Diffusion of phenol through a biofilm grown on activated carbon particles in a draft tube three-phase fluidized bed bioreactor. Biotechnol. Bioeng. 35:279-286

Fu Y-C, Zhang TC, Bishop PL (1994) Determination of effective oxygen diffusivity in biofilms grown in a completely mixed biodrum reactor. Wat. Sci. Tech. 29:455-462

Kitsos HM, Roberst RS, Jones WJ, Tornabene TG (1992) An experimental study of mass diffusion and reaction rate in an anaerobic biofilm. Biotechnol. Bioeng. 39:1141-1146

Lens P, De Beer D, Cronenberg CCH, Houwen FP, Ottengraf SPP, Verstraete WH (1993) heterogeneous distribution of microbial activity in methanogenic aggregates: pH and glucose microprofiles. Appl. Environ. Microb. 59:3803-3815

Libicki SB, Salmon PM, Robertson CR (1988) The effective diffusive permeability of a nonreacting solute in microbial cell aggregates. Biotechnol. Bioeng. 32: 68-85

Perry RH, Green PW eds.(1988) Perry's chemical engineers handbook, (6th ed.), McGraw-Hill Book Co., Singapore

Revsbech NP, Ward DM (1983) Oxygen microelectrode that is insensitive to medium chemical composition: Use in an acid microbial mat dominated by Cyanidium caldarum. Appl. Environm. Microbial. 45:755-759

Ruggeri B, Gianetto A, Sicardi S, Specchia V (1991) Diffusion phenomena in the spherical matrices used for cell immobilization. Chem. Eng. J. 46B:21-29

Tatevossian A (1985) The effect of heat inactivation, tortuosity, extracellular polyglucan and on-exchange sites of the diffusion of [^{14}C]-sucrose in human dental plaque residue in vitro. Arch. Oral Biol. 30: 365-371

Westrin BA, Axelsson A (1991) Diffusion in gels containing immobilized cells: A critical review. Biotechnol. Bioeng. 38:439-446

Westrin BA (1991) Diffusion measurements in gels: a methodological study. PhD-thesis, Lund University, Sweden

Westrin BA, Zacchi G (1991) Measurement of diffusion coefficients in gel beads:random and systematic errors. Chem. Eng. Sci. 46:1911-1916

▨ Abbreviations

A	surface of the membrane, [m^2]
C	concentration, [mol · m^{-3}]
D	diffusion coefficient, [m^2 · s^{-1}]
J	flux, [mol · m^{-2} · s^{-1}]
L	membrane thickness, [m]
m	partition coefficient, [-]
N	flow, [mol · s^{-1}]
t	time, [s]
R	radius, [m]
R	radial position, [m]
V	volume, [m^3]
β	time constant, [s]
ϕ	volume fraction, [-]
τ	tortuosity, [-]
app	apparent
c	continuous phase
d	dispersed phase
eff	effective

Quantity of Biomass Immobilized, Determination of Biomass Concentration

ELLEN A. MEIJER and RENÉ H. WIJFFELS

Introduction

In many immobilized-cell studies it is necessary to quantify the biomass in the gel. Biomass concentrations can be quantified by e.g. dry weight, cell number and protein measurements. A complicating factor in the case of immobilized cells is the presence of a matrix. For most methods biomass has to be separated from the immobilization matrix. This needs to be done completely without any loss of viability (when cell number or activity is determined) or destruction of the material (protein).

It is not possible to give general methods for determination of the concentration of immobilized biomass. The efficiency of the method will be dependent on the strength of the immobilization material and of the organism immobilized. Therefore, we chose to give a general approach to determine the concentration of immobilized biomass. Exact tuning needs to be done for each case. How this should be done has been described. In the method described, we use the protein content as an estimate for biomass concentration as an example.

Quantification of proteins of immobilized cells can be divided into three parts:

1. Isolation of cells from the gel matrix

2. Isolation of proteins from the cells

3. Quantification of those proteins

Ellen A. Meijer, WageningenUniversity, Food and Bioprocess Engineering Group, P.O. Box 8129, Wageningen, 6700 EV, The Netherlands

✉ René H. Wijffels, WageningenUniversity, Food and Bioprocess Engineering Group, P.O. Box 8129, Wageningen, 6700 EV, The Netherlands (*phone* +31-0317-484372; *fax* +31-0317-482237; *e-mail* rene.wijffels@algemeen.pk.wag-ur.nl)

The protein content of cells is dependent on many factors such as culture conditions (Park et al. 1990), cell type (Kocková-Kratochvílová 1990; Becker 1994) and growth phase. (Kocková-Kratochvílová 1990; Belkoura et al. 1997). The extraction of proteins from cells can be difficult due to a strong cell membrane or the presence of a cell wall or poor solubility of proteins like membrane bound proteins. The addition of low concentrations of detergents can overcome this problem (Wheelwright 1991; Bollag et al. 1996). However, the detergent used should not interfere with the dye used for the quantification. In addition, different quantification methods based on staining with a dye give different results (Berges 1993; Meijer and Wijffels 1998). Variations are caused by the fact that the dyes bind to different amino acids or bands of the proteins (Wiechelman et al. 1988; Wheelwright 1991). Also other cell components like sugars, lipids, free amino acids and nucleic acids, can interfere with the dyes (Bollag et al. 1996).

Considering these variations in dye staining and cell composition we have to realize that the use of fast extraction methods combined with protein dye quantification can give inaccurate results. Therefore when a method, developed and optimized for one cell type, is used for other cell types the obtained protein concentration has to be compared to the 'true' protein concentration, and if necessary the method has to be changed and optimized for the new cell type. For measurement of the 'true' protein concentration a Kjeldahl destruction and quantification of the reduced nitrogen is commonly used (Williams 1984). However, Kjeldahl destruction is an insensitive method requiring large amounts of sample. Therefore it is often not useful as protein quantification during experiments in which only small samples are available.

During the isolation of cells from the gel it is important that there are no changes in protein content of the cells and that cells are not damaged or lost. Protein extraction from immobilized cells has been described in some protocols (Freeman et al. 1982; Cheong et al. 1990; Smith et al. 1991). They all combined the extraction of cells from the gel and the extraction of proteins from the cells in 1 step.

Outline

It is impossible to develop a method for the extraction of proteins from all cell types immobilized in all gel materials. Therefore in this chapter a general framework to optimize the extraction and quantification method of proteins from immobilized cells will be presented. The method to obtain the protein concentration by Kjeldahl destruction will be discussed.

Furthermore one complete example for the quantification of baker's yeast proteins immobilized in κ-carrageenan will be shown in detail and compared to the protein concentration of the same batch yeast, but without immobilization.

Measurement of protein content by destruction to ammonium

With this method the nitrogen present in the cells is reduced to NH_3 by destruction of the cells in H_2SO_4 in the presence of a catalyst. Before destruction with boiling H_2SO_4 (above $300°C$) the water in the samples has to be removed by heating the sample at a lower temperature. This destruction takes several hours. As catalyst metals such as HgO, K_2SO_4 and $CuSO_4$ are used (Bailey 1967; Herbert et al. 1971; Williams 1984; Greenberg et al. 1985).

After this destruction the NH_3-N concentration can be measured by titration after distillation of the ammonia into HCl or boric acid (Bailey 1967; Herbert et al. 1971; Williams 1984; Greenberg et al. 1985). More sensitive quantification methods are colorimetric reactions, like nesslerization and the phenate-method. By nesslerization the ammonia forms a complex with Hg and I, that can be measured quantitatively at wavelengths between 400 and 500 nm (Bailey 1967; Greenberg et al. 1985). With the phenate method ammonia, active chlorine and phenol form indophenol. This is a colored compound whose color can be intensified by the addition of nitroprusside. Measuring the absorbance at 660 nm gives the amount of NH_3-N (Greenberg et al. 1985). Eventually this last method can be automated.

The protein concentration can be calculated by multiplying the NH_3-N concentration (in mg/l) with a factor 6.25 (Herbert et al. 1971; Williams 1984; Becker 1994).

General setup for the development of a valid protocol for the quantification of proteins in cells immobilized in a gel

Step 1. Development or control of the extraction of proteins from free cells

Be sure that the method to measure the protein content of free cells is effective. As already mentioned the protein content of cells is dependent on cell type and growth conditions. Therefore, for the development of a method it is important to relate the obtained protein content by the method used to the true protein content measured by Kjeldahl destruction.

Rapid extraction methods have been described for almost all cell types. For example, yeast proteins can be extracted by boiling in a OH⁻ solution (Verduyn et al. 1990) or by sonication (Park et al. 1990), proteins of micro algae are effectively isolated by sonication (Meijer and Wijffels 1998). Often no comparison is made to the real protein content, resulting in an under-estimation of the protein concentrations (Rausch 1981). Meijer and Wijffels (1998) have shown that a method developed for one species can not be used without any control for other species. In Table 1 an effective sonication method to extract proteins from *Chlorella* was used to extract proteins from other algal species. These protein concentrations are expressed as % of the protein concentration measured by Kjeldahl-destruction. For other algae species recoveries varied from 4 to 86 %.

Table 1. Efficiency of a sonication method, developed for *Chlorella*, compared to Kjeldahl destruction for several algae (Meijer and Wijffels 1998)

Algae	efficiency (%)
Chlamydomonas	86
Chlorella	103
Dunaliella	10
Rhodomonas	4
Synechococcus	50

Step 2. Dissolving the beads

Make beads without cells in it. Develop a method to dissolve the beads. Depending on the type of gel used, different methods to dissolve the gel material are possible.

1. Heating the gel followed by dilution (Smith et al. 1991). Cells might be damaged too, so it can be combined with the extraction of the proteins out of the cells, for example by boiling the beads in 1 M NaOH (Freeman et al. 1982; Smith et al. 1991).

2. Sonication of the gel. Since sonication is also used to break the cell membrane and cell wall of cells, cells might be damaged. Combination of the breakdown of the gel and extraction of the protein is then very useful. This sonication of the gel is expected to be very effective for synthetic gels like poly ethylene glycol and poly vinyl alcohol.

3. Removing of the stabilizing ions. This is a very mild method, but either time-consuming or resulting in large dilution factors. The cells will stay intact, but when it goes very slowly cell composition will change. This change results in a protein measurement that is not related to the time the sample was taken. When the cells are diluted too much, protein concentrations can become too low to be measured.

Step 3. Interference of gel at protein staining

Test the gel material and the solutions in which the beads will be grown/cultured and dissolved, for influences on the staining with the dye. If staining of this solution is too strong, use other buffers or another protein dye.

Step 4. Measurement of the protein concentration of immobilized cells and comparison of this concentration with the protein concentration of free cells

Prepare cells for immobilization. Separate a fraction as a control. Immobilize the rest of the cells and bring the beads in their culture conditions, but do not culture them. Directly use the beads to dissolve and to extract the proteins out of the cells. Pay special attention to possible swelling or shrinking of the gel since this will lead to unnoticed concentration or dilution. As control extract the control cells in exactly the same way. Control that the immobilized cells give the same protein content as the free cells. For the quantification of the proteins fast, sensitive and easy to use dyes are available such as Coomassie Brilliant Blue (Bradford 1976), Folin reagent (Lowry 1951) and bicinchoninic acid (Smith et al. 1985). The staining with these dyes is dependent on the amino acid composition of the samples and thereby can result in different protein concentrations (Berges et al. 1993; Meijer and Wijffels 1998).

Materials

Step by step protocol for yeast proteins in κ-carrageenan

The following protocol was developed for yeast cells immobilized in κ-carrageenan beads with an average bead diameter of 2 mm. Generally beads show some variation in size, so to measure a reliable protein concentration the assay has to be performed at least in triplicate with a minimal 10 beads per assay to exclude effects of variations in volume of the sample.

Equipment and reagents

– Spectrophotometer

– *Saccharomyces cerevisiae*, immobilized in 2% κ-carrageenan beads.

– Coomassie Brilliant Blue: dissolve 100 mg Coomassie Brilliant Blue G-250: in 50 ml 95 %(v/v) ethanol, add 100 ml 85% (w/v) phosphoric acid and fill up to 1 liter with demineralized water.

– PBS: Dissolve 8.2 g NaCl, 1.9 g $Na_2HPO_4.2H_2O$ and 0.3 g $NaH_2PO_4 \cdot 2H_2O$ in 1 liter demineralized water. pH should be 6.8.

– 0.1 M NaOH

– 4 M HCl

– BSA 0 to 100μg/ml dissolved in water.

Procedure

Isolation

1. Wash the beads in PBS 15 minutes at room temperature. Use therefore 5 to 10 times more PBS than the volume of the beads.

2. Remove PBS from the beads.

3. Add 10 bead volumes 0.1 M NaOH.

4. Put samples 30 minutes at 100°C.

5. While the sample is still hot add 0.25 volume 4 M HCl and mix. This neutralization is important for the following protein staining which is performed in acidic medium.

6. Dilute 10 times in water immediately. After this dilution the concentrations of K^+ and κ-carrageenan are so low that the sample will not gel again.

Quantification

Always check that the remaining medium in which the beads have been cultured does not influence the staining of the Coomassie Brilliant Blue to avoid false interpretations of the results.

1. Depending on protein concentration dilute the samples in water.

2. Add 1 ml reagent in the cuvettes.

3. Add 100 μl BSA or sample.

4. After 2 minutes to 1 hour at room temperature measure A_{595}.

5. Protein content can be calculated by the use of standards.

Results

The following obtained result is only an illustration of the efficiency of the extraction. The absolute concentrations found are dependent on number of cells in the beads, cell type and culture conditions. By using the described method the protein concentration of freshly immobilized *Saccharomyces cerevisiae* was 169.2 ± 31.9 µg/ml. When the same yeast cells were treated comparable to the immobilized cells, but without immobilization a protein concentration of 194.9 ± 13.9 µg protein / ml buffer was found. Empty beads gave a protein concentration of -0.5µg/ml, which is negligible to the measured concentrations.

Troubleshooting

- When samples gel again before adding HCl and following dilution, heat sample to 70 °C and add directly the HCl and dissolve.

- When samples still gel after dilution, perform that dilution in warm buffer and do a subsequent dilution step.

- When the protein concentration is out of the BSA range, dilute samples more.

- When you cannot detect any protein consider the more sensitive method of Bearden, in which not only the formation of the blue color is measured but also the disappearance of the free dye (Bearden 1978). If this does not help, try to use less NaOH to extract the proteins. Or consider another quantification method like cell counting (Martins dos Santos et al. 1997) or oxygen uptake (Wijffels et al. 1991)

References

Bailey JL (1967) Techniques in protein chemistry. Elsevier publishing company, Amsterdam London New York

Bearden jr. JC (1978) Quantitation of submicrogram quantities of protein by an improved protein-dye binding assay. Biochim Biophys Acta 533:525-529

Becker EW (1994) Micoalgae: biotechnology and microbiology. Cambridge university press, Cambridge

Belkoura M, Benider A, Dauta A (1997) Influence de la température, de l'intensité lumineuse et du stade de croissance sur la composition biochimique de *Chlorella sorokiniana* Shihira & Krauss. Annls Limnol 1:3-11

Berges JA, Fisher AE, Harrison PJ (1993) A comparison of Lowry, Bradford and Smith protein assays using different protein standards and protein isolated from the marine diatom *Thalassoria pseudonana*. Marine Biology 115:187-193

Bollag DM, Rozycki MD, Edelstein SJ (1996) Protein methods. Second edition. Wiley Liss, new York Chichester Brisbane Toronto Singapore

Bradford MM (1976) A rapid and sensitive method for the quantitation of microgram quantities of protein utilizing the principle of protein-dye binding. Anal Biochem 72:248-254

Cheong KH, Katayama Y, Seto M, Kuraishi H (1990) Estimation of cellular protein for monitoring the cell growth of bacteria in a photo-crosslinked polymer resin. J Ferm Bioeng 70:136-138

Freeman A, Blank T, Aharonowitz Y (1982) Protein determination of cells immobilized in cross-linked synthetic gels. Eur J Appl Microbiol Biotechnol 14:13-15

Greenberg AE, Trussell RR, Clesceri LS (1985) Standard methods for the examination of water and wastewater, 16th edition. American Public Health Association, Washington DC

Herbert D, Phipps PJ, Strange RE (1971) Chemical analysis of microbial cells. In: Norris JR (ed) Methods in microbiology, vol 5B. Academic Press, London New York pp 210-344

Kocková-Kratochvílová (1990) Yeasts and yeast-like organisms. VCH publishers, Weinheim New York Cambridge Basel

Lowry OH, Rosebrough NJ, Farr AL, Randall RJ (1951) Protein measurement with the Folin phenol reagent. J Biol Chem 193:265-275

Martins dos Santos VAP, Vasilevska T, Kajuk B, Tramper J, Wijffels RH (1997) Production and characterization of double-layer beads for coimmobilization of microbial cells. Biotechnol Ann Rev 3:227-244

Meijer EA, Wijffels RH (1998) Development of a fast, reproducible and effective method for the extraction and quantification of proteins of micro algae. Biotechnol Tech 12:353-358

Park WS, Murphy PA, Glatz BA (1990) Lipid metabolism and cell composition of the oleaginous yeast *Apiotrichum curvatum* grown at different carbon to nitrogen ratios. Can J Microbiol 36:318-326

Rausch T (1981) The estimation of micro-algal protein content and its meaning to the evaluation of algal biomass I. Comparison of methods for extracting protein. Hydrobiologia 78:237-251

Smith MR, Haan A de, Wijffels RH, Beuling EE, Vilchez C, Bont JAM de (1991) Analysis of growth and specific activities of immobilized microbial cells. Biotechnol Tech 5:323-326

Smith PK, Krohn RI, Hermanson GT, Mallia AK, Gartner FH, Provenzano MD, Fujimoto EK, Goeke NM, Olson BJ, Klenk DC (1985) Measurement of protein using bicinchoninic acid. Anal Biochem 150:76-85

Verduyn C, Postma E, Scheffers WA, Dijken JP van (1990) Physiology of *Saccharomyces cerevisiae* in aerobic glucose-limited chemostat cultures. J General Microbiol 136:395-403

Wheelwright SM (1991) Protein purification: design and scale up of downstream processing. Hanser publishers, Munich Vienna New York Barcelona

Wiechelman KJ, Braun RD, Fitzpatrick JD (1988) Investigation of the bicinchoninic acid protein assay: identification of the groups responsible for color formation. Anal Biochem 175:231-237

Wijffels RH, Gooijer CD de, Kortekaas S, Tramper J (1991) Growth and substrate consumption of *Nitrobacter agilis* cells immobilized in carrageenan: part 2. Model evaluation. Biotechnol Bioeng 38:232-240

Williams (1984) Official methods of analysis of the Association of Official Analytical Chemists, 14th edition. Association of Official Analytical Chemists Inc, Arlington

General information dealing with protein quantification, cell composition and protein synthesis in yeasts and algae can be found in the following books. Those are only a few examples of many books published about these subjects.

Becker EW (1994) Micoalgae: biotechnology and microbiology. Cambridge university press, Cambridge

Bollag DM, Rozycki MD, Edelstein SJ (1996) Protein methods. Second edition. Wiley Liss, new York Chichester Brisbane Toronto Singapore

Kocková-Kratochvílová (1990) Yeasts and yeast-like organisms. VCH publishers, Weinheim New York Cambridge Basel

Wheelwright SM (1991) Protein purification: design and scale up of downstream processing. Hanser publishers, Munich Vienna New York Barcelona

Kinetics of the Suspended Cells

RENÉ H. WIJFFELS

Introduction

Micro-organisms which are entrapped in gel beads grow, consume substrate and produce products in a similar way to suspended cells. Immobilization by entrapment is often so gentle that the immobilized cells have intrinsically the same kinetics as the suspended cells would have. Observed differences are generally caused by diffusion limitation. Substrate has to be transported from the liquid medium surrounding the solid particles and the particles and products have to be transported in the opposite direction. As a consequence gradients in concentration of substrates and products are present and local concentrations in the beads are different from the liquid medium. As a result the kinetics of immobilized cells is apparently different from suspended cells. These differences can in most cases be contributed to mass transfer phenomena.

Models have been developed in which cell kinetics and mass transfer have been integrated. Based on the kinetics of suspended cells and mass transfer equations the performance of the immobilized cell process can be predicted very well.

Outline

The intrinsic kinetics of immobilized cells is thus different from the kinetics of suspended cells. In order to determine the kinetics without having influence of diffusion limitation kinetic parameters are determined with suspended cells. In this chapter it will briefly be explained how to deter-

René H. Wijffels, WageningenUniversity, Food and Bioprocess Engineering Group, P.O. Box 8129, Wageningen, 6700 EV, The Netherlands (*phone* +31-0317-484372; *fax* +31-0317-482237; *e-mail* rene.wijffels@algemeen.pk.wag-ur.nl l)

mine suspended cell kinetics. As methods for this are so general the phenomena will be introduced and for detailed description of the methods other works will be referred to.

Growth models

In order to determine the kinetics of suspended cells, some kinetic equations should be given. The equation for growth rate is:

$$r_x = \mu X$$

With:

$$\mu = \mu_{max} \frac{S}{K_s + S}$$

The equation for substrate consumption is:

$$r_S = \frac{r_x}{Y_{xs}} + m_{sx}$$

Procedure

The kinetic parameters of the suspended cells, μ_{max}, K_s, Y_{xs} and m_s, can be determined both in batch experiments and continuous culture experiments.

In general, cultivations are done and substrate and biomass concentrations are monitored in time (in case of batch experiments) or at different dilution rates (in continuous experiments). With the help of the mass balances for substrate and product in which the equations given above have been substitutes the kinetic parameters can be obtained.

Biomass balance: Batch reactor

$$\frac{dX}{dt} = r_x$$

Substrate balance:

$$\frac{dS}{dt} = r_s S$$

Continuous reactor

Biomass balance:

$$\mu = D$$

Substrate balance:

$$D(S_{in} - S) = r_s$$

For further details refer to the literature given in 'References' where procedures have been described.

References

Atkinson B., Mavituna F. (1983) Biochemical engineering and biotechnology handbook. Stockton Press, p 1271

Doran P.M. (1995) Bioprocess engineering principles. Academic Press, p 439

Pirt S.J. (1975) Principles of microbe and cell cultivation. Blackwell scientific publications, p 274

Schlegel H.G. (1986) General microbiology. Cambridge University Press, p 587

Stephanopoulos G. (1993) Bioprocessing. Vol. 3 of Biotechnology (eds. H.-J. Rehm, G. Reed, A. Pühler, P. Stadler), p 816

Van 't Riet, Tramper (1991) Basic bioreactor design. Marcel Dekker, Inc., p 465

Abbreviations

D	dilution rate (s^{-1})
K_s	Monod constant ($mol \cdot m^{-3}$)
m_s	maintenance coefficient ($mol \cdot kg^{-1} \cdot s^{-1}$)
S	substrate concentration ($mol \cdot m^{-3}$)
S_{in}	substrate concentration of the medium with which the reactor is fed ($mol \cdot m^{-3}$)
r_s	volumetric rate of substrate consumption ($mol \cdot m^{-3} \cdot s^{-1}$)
r_x	volumetric rate of biomass production ($kg \cdot m^{-3} \cdot s^{-1}$)
t	time (s)
X	viable biomass concentration ($kg \cdot m^{-3}$)
Y_{xs}	true yield of biomass from substrate (kg biomass/mol substrate)
μ	specific growth rate (s^{-1})
μ_{max}	maximum specific growth rate (s^{-1})

Diffusion Limitation

RENÉ H. WIJFFELS

Introduction

Numerous papers have been devoted to modelling simultaneous diffusion and conversion of substrate by immobilized biocatalysts. In such systems the substrate is transferred from a liquid phase to a solid phase in which the reaction occurs. In many of these studies the rate-limiting step in the process is the diffusion of substrate through the solid phase.

In the case of immobilized cells not only diffusion and consumption of substrate occurs simultaneously, but growth and decay of biomass takes place as well. Several dynamic models have been developed accounting for all these processes (De Gooijer et al. 1991, Wijffels et al. 1991, 1994, 1995, 1996 Wijffels and Tramper 1995, Dos Santos et al. 1996, Leenen et al. 1997, Monbouquette et al. 1991, Hunik et al. 1994, Willaert et al. 1995).

For non-growing biocatalysts the set of equations is relatively simple to solve. In case growth occurs the situation is more complex. In this chapter it is assumed that growth does not occur and that the biomass is homogeneously distributed across the support material. The aim of this chapter is not to present those dynamic models. A set of equations is given allowing relatively simple calculations to determine to what extent diffusion limitation is of importance. In case diffusion limitation is important and you wish to understand the process better it is advised to read further on this subject.

In the cases where the immobilization material is fully grown with cells; this means when the maximum biomass concentration is reached the calculations given in this chapter are very useful. In that case we do not have to consider growth of cells resulting in a change in biomass concentration.

René H. Wijffels, WageningenUniversity, Food and Bioprocess Engineering Group, P.O. Box 8129, Wageningen, 6700 EV, The Netherlands (*phone* +31-0317-484372; *fax* +31-0317-482237; *e-mail* rene.wijffels@algemeen.pk.wag-ur.nl)

In these cases cell growth results in release of biomass from the beads and the immobilized biomass concentration remains constant (Wijffels et al. 1994).

Outline

The phenomena of diffusion limitation are described. Both internal and external diffusion limitation are calculated. Calculations will be done for zero and first order substrate consumption kinetics. In these cases equations can be solved analytically. If substrate consumption is in between (Monod type of kinetics) diffusion limitation can only be calculated by means of numerical solutions. A method is presented to estimate diffusion limitation with analytical methods as well.

Theory   Immobilized cells consume substrate. The substrate consumed needs to be transported in the support in which the cells are immobilized. Transport of the substrate is assumed to take place via diffusion. Per definition, the rate of substrate consumption is equal to the rate with which the substrate is consumed. The overall rate can, however, be determined by the rate of conversion or by the rate of diffusion. In the latter case a gradient in substrate concentration develops across the support and the process is 'diffusion limited'.

Gradients can be present both across the particle itself (internal diffusion limitation) and across a stagnant liquid layer surrounding the particle (external diffusion limitation). As the result of a stagnant layer surrounding the biocatalyst particle (the bulk is assumed to be ideally mixed) the concentration of the substrate at the surface of the particle is lower than in the bulk of the liquid.

Diffusion limitation can be quantified with an effectiveness factor (η). The effectiveness factor is defined as the observed conversion rate (so under the circumstances of diffusion limitation) divided by the conversion rate that would be observed if there was no diffusion limitation. In general the effectiveness factor is between 0 and 1. In case $\eta=1$, there is no diffusion limitation.

The effectiveness factor can be substituted in the equation for substrate consumption (disregarding maintenance):

$$r'_s = \eta \frac{\mu_{max}X}{Y_{xs}} \frac{S_b}{K_s + S_b} \tag{1}$$

The effectiveness factor (η) consists of an external effectiveness factor (η_e) and an internal effectiveness factor (η_i). The overall effectiveness factor is the product of internal and external effectiveness factor:

$$\eta = \eta_e \cdot \eta_i \tag{2}$$

The external effectiveness factor is defined as the conversion rate that would be observed at the substrate concentration of the liquid/solid interface of the particle divided by the conversion rate observed if there was no diffusion limitation. The internal effectiveness factor is defined as the observed conversion rate divided by the conversion rate that would be observed at the substrate concentration of the liquid/solid interface of the particle.

The external effectiveness factor can be calculated by (Van 't Riet and Tramper 1991):

External effectiveness factor

$$\eta_e = \frac{S_i(K_s + S_b)}{S_b(K_s + S_i)} \tag{3}$$

In general all parameters in this equation are known except for the interfacial substrate concentration S_i.

The interfacial concentration S_i can be calculated by the balance equation (consumption = transport):

$$\eta_i \frac{\mu_{max} X}{Y_{xs}} \frac{S_i}{K_s + S_i} \varepsilon_p = k_{il} \dot{A}(S_b - S_i) \tag{4}$$

This equation says that the amount of substrate that is consumed should also be transported across the stagnant layer.

In this equation:

$$\dot{A} = \frac{6}{d_p} \varepsilon_p \tag{5}$$

How the internal effectiveness factor can be calculated will be shown later. How the liquid-solid mass transfer coefficient (k_{il}) can be calculated has been described in Chapter 16.

With equation (4) the interfacial substrate concentration (S_i) can be calculated.

The internal effectiveness factor is calculated via the Thiele modulus. The Thiele modulus is the ratio of the kinetic rate and the diffusion rate. Calculations will be shown for zero order kinetics and first order kinetics for immobilized cells as for those situations analytical solutions can be used. Further backgrounds are given in Van 't Riet and Tramper (1991).

Internal effectiveness factor

The shape of the support material is spherical and the biomass is homogeneously distributed over the support material.

Zero order kinetics

The Thiele modulus for zero order kinetics is (Van 't Riet and Tramper 1991):

$$\phi^2 = \frac{\mu_{max}X}{Y_{xs}} \frac{R_p^2}{18 D_{sp} S_i} \tag{6}$$

With the Thiele modulus the effectiveness factor can be calculated:

If $\phi \leq 0.57$ then $\eta_{i0} = 1$

If $\phi > 0.57$ then :

$$\eta_{i0} = 1 - x^3 \tag{7}$$

and:

$$x = cos\left(\frac{y + 4\pi}{3}\right) + \frac{1}{2} \quad \text{(to be calculated in radials)} \tag{8}$$

and:

$$y = cos^{-1}\left(\frac{2}{3\phi^2} - 1\right) \quad \text{(to be calculated in radials)} \tag{9}$$

First order kinetics

The Thiele modulus for first order kinetics is (Van 't Riet and Tramper 1991):

$$\phi = \frac{1}{6} d_p \sqrt{\frac{\mu_{max}X}{Y_{xs}K_s D_{sp}}} \tag{10}$$

With the Thiele modulus the effectiveness factor can be calculated:

$$\eta_{i1} = \frac{3\phi coth(3\phi) - 1}{3\phi^2} \tag{11}$$

with:

$$coth(3\phi) = \frac{e^{3\phi} + e^{-3\phi}}{e^{3\phi} - e^{-3\phi}} \tag{12}$$

Monod kinetics
In case zero or first order kinetics are oversimplifications and Monod kinetics applies, the effectiveness factor can be estimated with a weighing factor (Van 't Riet and Tramper 1991):

$$\theta = \frac{S_i}{K_s + S_i} \tag{13}$$

$$\eta_i = \theta\eta_{i0} + (1 - \theta)\eta_{i1} \tag{14}$$

Procedure

With the equations given above the effectiveness factor can be calculated directly. In case internal and external diffusion limitation occurs, the calculation procedure is slightly more complicated as the external effectiveness factor can only be calculated when the internal effectiveness factor is known and for the internal effectiveness factor the interfacial substrate concentration is needed. In these cases an iteration procedure is necessary. This procedure is given below, in which is reference is made to the calculations given above.

1. In order to calculate the internal effectiveness factor the interfacial substrate concentration should be known. As this concentration is not known, the iteration procedure is started with the assumption that the interfacial concentration is equal to the substrate concentration in the bulk liquid ($S_i = S_b$)

2. Calculate the zero order Thiele modulus with equation (6)

3. If $\phi \leq 0.57$ then $\eta_{i0} = 1$

4. If $\phi > 0.57$ then calculate η_{i0} with equations (7-9)

5. Calculate the first order Thiele modulus with equation (10)

6. Calculate η_{i1} with equations (11) and (12)

7. Calculate θ with equation (13)

8. Calculate the internal effectiveness factor (η_i) with equation (14)

9. Calculate the external mass transfer coefficient (k_{il}) as has been described in Chapter 16

10. Substitute k_{il} and the calculated value of η_i in equation (4) and calculate S_i

11. If S_i is smaller than the value assumed under point 1 the calculation procedure should be repeated with the new value of S_i. This iteration procedure should be continued until the calculated value of S_i is equal to the value assumed. During the iteration procedure steps 5, 6 and 9 do not need to be repeated as these values are independent of the interfacial concentration.

12. Calculate the external effectiveness factor with S_i and equation (3)

13. Calculate the overall effectiveness factor with equation (2)

Troubleshooting

The procedure given above is a simplified procedure. The effectiveness factor is calculated for steady state situation, i.e. it is assumed that the biomass concentration does not change. In reality the immobilized cell process is a very dynamic process. Nevertheless, the above approach gives insight into what extent diffusion limitation is important in your process. If a detailed and exact description of the process is needed, dynamic models need to be used. For these dynamic models is referred to literature (De Gooijer et al. 1991, Wijffels et al. 1991, 1994, 1995, 1996 Wijffels and Tramper 1995, Dos Santos et al. 1996, Leenen et al. 1997, Monbouquette et al. 1991, Hunik et al. 1994, Willaert et al. 1995).

The assumptions used in the procedure presented are summarized:

1. Spherical particles are considered.

2. Calculations are done for zero and first order kinetics and for Monod kinetics estimations are given.

3. The immobilized biomass concentration is constant in time.

4. Biomass is homogeneously distributed across the support material.

References

Doran P.M. (1995) Bioprocess engineering principles. Academic Press, 439 p

Dos Santos V.A.P.M., Marchal L.M., Tramper J., Wijffels R.H. (1996) Modeling and evaluation of an integrated nitrogen removal system with microorganisms co-immobilized in double-layer gel beads. Biotechnol. Prog. 12, 240-248

De Gooijer C.D., Wijffels R.H., Tramper J. (1991). Growth and substrate consumption of *Nitrobacter agilis* cells immobilized in carrageenan. Part 1: Dynamic modeling. Biotechnol. Bioeng. 38: 224-231

Hunik J.H., Bos C.G., Van den Hoogen M.P., de Gooijer C.D., Tramper J. (1994) Co-immobilized *Nitrosomonas europaea* and *Nitrobacter agilis* cells: validation of a dynamic model for simultaneous substrate conversion and growth in <K>-carrageenan gel beads. Biotechnol. Bioeng. 43: 1153-1163

Leenen E.J.T.M., Boogert A.A., Van Lammeren A.A.M., Tramper J., Wijffels R.H. (1997) Dynamics of artificially immobilized *Nitrosomonas europaea*: effect of biomass decay. Biotechnol. Bioeng. 55: 630-641

Monbouquette H.G., Sayles G.D., Ollis D.F. (1990) Immobilized cell biocatalyst activation and pseudo-steady state behaviour: model and experiment. Biotechnol. Bioeng. 35: 609-629

Van 't Riet K. & Tramper L. (1991) Basic Bioreactor Design. Marcel Dekker Inc.

Van 't Riet K. & Tramper L. (1991) Basic Bioreactor Design. Marcel Dekker Inc.

Wijffels R.H., de Gooijer C.D., Kortekaas S., Tramper J. (1991). Growth and substrate consumption of *Nitrobacter agilis* cells immobilized in carrageenan. Part 2: Model evaluation. Biotechnol. Bioeng. 38: 232-240

Willaert R.G., Baron G.V., De Backer L. (1995) Immobilised living cells systems: modelling and experimental methods. John Wiley, 373 p.

Wijffels R.H., De Gooijer C.D., Schepers A.W., Beuling E.E., Mallée L.R., Tramper J. (1995) Growth of immobilized *Nitrosomonas europaea*: implementation of diffusion limitation over microcolonies. Enzyme and Microbial Technology 17: 462-471

Wijffels R.H., Schepers A.W., Smit M., De Gooijer C.D., Tramper J. (1994) Effect of initial biomass concentration on the growth of immobilized *Nitrosomonas europaea*. Appl Microbiol Biotechnol. 42: 153-157

Wijffels R.H., Tramper J. (1995) Nitrification by immobilized cells. Enzyme and Microbial Technology 17: 482-492

Wijffels R.H., Buitelaar R.M., Bucke C., Tramper J. (1996) Immobilized cells: basics and applications. Progress in Biotechnology vol. 11, Elsevier, 845 p.

Wijffels R.H., Buitelaar R.M., Bucke C., Tramper J. (1996) Immobilized cells: basics and applications. Progress in Biotechnology vol. 11, Elsevier, 845 p.

Wijffels R.H., Tramper J. (1995) Nitrification by immobilized cells. Enzyme and Microbial Technology 17: 482-492

Willaert R.G., Baron G.V., De Backer L. (1995) Immobilised living cells systems: modelling and experimental methods. John Wiley, 373 p.

Abbreviations

A'	specific surface area of the particle per unit reactor volume (-)
d_p	particle diameter (m)
D_{sp}	diffusion coefficient of the substrate in the particle ($m^2 \cdot s^{-1}$)
k_{il}	liquid-solid mass transfer coefficient for component i ($m \cdot s^{-1}$)
K_s	Monod constant ($mol \cdot m^{-3}$)
R_p	particle radius (m)
r_s'	observed (apparent) volumetric rate of substrate consumption ($mol \cdot m^{-3} \cdot s^{-1}$)
S_b	substrate concentration in the bulk phase ($mol \cdot m^{-3}$)
S_i	substrate concentration at the liquid/solid interphase ($mol \cdot m^{-3}$)
X	viable immobilized biomass concentration (per volume of particle) ($kg \cdot m^{-3}$)
Y_{xs}	true yield of biomass from substrate (kg biomass/mol substrate)
μ_{max}	maximum specific growth rate (s^{-1})
η	effectiveness factor (-)
η_e	external effectiveness factor (-)
η_i	internal effectiveness factor (-)
η_{i0}	internal effectiveness factor for zero order kinetics (-)
η_{i1}	internal effectiveness factor for first order kinetics (-)
θ	weighing factor (-)
ε_p	particle hold-up (volume of particles per volume of reactor) (-)
ϕ	Thiele modulus (-)

Micro-Electrodes

DIRK DE BEER

▦ Introduction

Microsensors are powerful tools for the determination of local fluxes and the distribution of microbial activity in sediments, microbial mats, biofilms, aggregates and other immobilized cell systems (ICS). Within these structures free convection is hindered, and consequently mass transfer to the cells often limits conversion rates. Determination of the microbial community using microbiological or molecular methods have obvious limitations for predictions of the behavior of the ICS. Firstly, any enumeration method, both the cultivation dependent and molecular methods have biases. Prediction of the behavior of immobilized biomass based on these enumerations are further biased by the unknown species distribution within the ICS. Because of mass transfer resistance the microenvironment in the ICS differs from the bulk medium. Consequently, extrapolation of the system behavior to that of the cells is impossible without knowledge about their microenvironment. Therefore, detection techniques with high spatial resolution are needed, both for microbial species and microbial activity distribution. New tools are molecular analyses of the genetic material, avoiding the bias of cultivation methods, and the use of microelectrodes to determine the chemical composition inside intact ICS.

A variety of microsensors relevant for microbial ecology have been developed and used, such as for O_2 (Revsbech, 1989; Revsbech and Ward, 1983), N_2O (Revsbech et al., 1988), pH (Hinke, 1969), NH_4^+ (De Beer and Van den Heuvel, 1988), NO_3^- (De Beer and Sweerts, 1989; Larsen et al., 1997; Larsen et al., 1996), S^{2-} (Revsbech et al., 1983), H_2S (Jeroschewski et al., 1996), NO_2^- (De Beer et al., 1997b), CH_4 (Damgaard and Revsbech,

Dirk de Beer, Max Planck Institute for Marine Microbiology, Celsiusstrasse 1, Bremen, 28359, Germany (*phone* +49-421-2028802; *fax* +49-421-2028690; *e-mail* dbeer@mpi-bremen.de)

1997), Ca^{2+} (Ammann et al., 1981) and CO_2 (De Beer et al., 1997a). With these sensors a wide variety of microbial processes has been studied, such as aerobic respiration (De Beer, 1990; Revsbech and Jørgensen, 1986; Revsbech et al., 1986; Sweerts et al., 1991), photosynthesis (Revsbech et al., 1983; Revsbech et al., 1981), nitrification (De Beer, 1990; De Beer et al., 1991; De Beer et al., 1993; Jensen et al., 1993; Sweerts and de Beer, 1989; Wijffels et al., 1995), denitrification (Jensen et al., 1993; Jensen et al., 1994), fermentation (De Boer et al., 1993), sulfate reduction (Kühl and Jørgensen, 1992; Kühl et al., 1998), sulfide oxidation (Kühl and Jørgensen, 1992; Kühl et al., 1998), methanogenesis (De Beer, 1990; Lens et al., 1995; Lens et al., 1994) and calcification (Hartley et al., 1996; McConnaughey and Falk, 1991).

Outline

This chapter will describe the manufacturing of the O_2 microsensor, which is probably the most important sensor and certainly the most widely used. Also the use of the sensor will be explained, i.e. details on the calibration, the measurement and calculation of microprofiles. It will be explained how microprofiles can be used to calculate distributions of activities. Finally, an example of a measurement will be used to clarify the different steps in the interpretation of the microprofiles.

Materials

Microsensor manufacturing

- microscope
- dissection scope
- heating loops: 1 mm Tantalum, 50 μm platinum wire.
- power source (Vario transformers, 0-20 V)
- micromanipulator
- watchmakers forceps
- Platinum wire, 50 μm, 99.99% pure, non-crystalline
- silver wire, 50 μm
- Bunsen burner, welding torch

– glass tubing

– Pasteur pipettes

– AR-glass, 5 mm OD, Schott 8418

– alkaline resistant glass: Green Schott, 3.5 mm OD, 8533, 8512 or 8516

– silicone, any clear silicone will work

– epoxy resin, two-component, curing within 30 minutes at room temperature

– Gold solution. This is made by adding a microspoon tip of gold-dust in 1 ml of *Aqua regia* (18% HNO_3, 82% concentrated HCl). This is left open for a few days until dried out. Then a drop of water is added to dissolve the salt. It can be used for years.

– micromanipulator (preferably motor-driven) mounted on heavy stand

– dissection microscope

– picoamperemeter

– recorder or data-acquisition equipment

– nitrogen gas

Microsensor measurement

Subprotocol 1
Manufacturing of O_2 Microsensors

▨▨ Procedure

1.1 Cathode

The cathode is made from a 5 cm piece of 50 μm thick Pt-wire. This is first etched in saturated KCN with 2 V AC applied between the Pt-wire and a graphite rod. The Pt-wire must be moved up and down during the etching in the KCN solution over a distance of a few cm. This results in a Pt-wire, tapered to a ca 1μm tip. The wire is rinsed with water, HCl, water and ethanol.

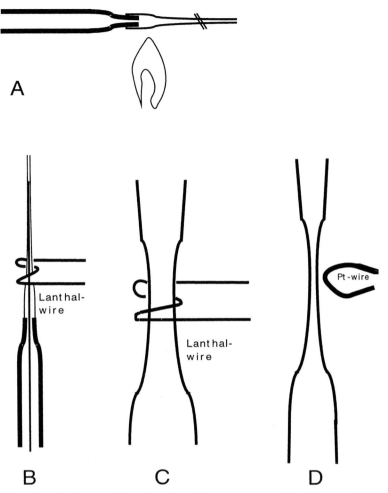

Fig. 1. Steps in the manufacturing of O_2 microsensors. **A** Fusing of green and white glass. **B** Melting of the tapered Pt-wire in green glass capillary. **C** Tapering of casing with large heating loop. **D** Tapering of casing to microtip by thin Pt-wire

Heat green glass tubing in a flame from a Bunsen burner and pull to a thickness of ca 1 mm over a length of ca 20 cm; turn glass between the fingers while pulling and cooling to keep straight. Cut 4 mm before the shoulder at the thick part. Pull white glass over a flame to a thickness of ca 2 mm over a length of 1 cm (take a piece of ca 15 cm, heat in the middle). Cut in two at the 'waist'. Then put the green piece over the tapered white glass (should fit closely), and fuse in a small hot flame (Fig. 1A). Turn along axis while sealing and heating. Both the green and the white glass must have been red-hot to obtain a good seal. The Pt-wire is inserted in the capillary and fused in the large heating loop. Hang the capillary in the big heating loop (thin green tip upwards), start heating just above the shoulder and heat until glass slowly elongates (no weights are needed) and seals around the Pt-wire (Fig. 1B). It will finally fall off. Observe this by a binocular.

The Pt-wire is now covered with glass and must be exposed. This is done by cutting the glass coating off using the watchmakers forceps. Ca 10 μm of Pt wire must be exposed. If the wire is very thin and a thin cathode is desired, cut the glass just above the Pt-tip and melt excess glass back with the small Pt-heating loop.

The cathode must be coated with gold by electroplating. This solution is brought in a Pasteur pipette and under microscopic observation the Pt-cathode is brought in. Then a potential of -0.6 V is applied (cathode negative) until a gold crust is formed on the tip with a diameter of ca 10 μm.

1.2 Casing

The outer casing is made from a Pasteur pipette. Pasteur pipettes are first pulled in a flame to a thickness of ca 0.5 mm over a length of 3-6 cm. The pipettes are clamped with tip upwards in the big heating loop, heated and pulled by gravity to a thickness of 200 μm (Fig. 1C). Then use the small heating loop to reduce the thickness further (Fig. 1D). The thinner the capillary, the further the heat must be reduced. Finally, the heating-loop is moved at low heat 'through' the capillary and it falls of. The tip is carefully broken to the desired size (ca 5-10μm). Then the tip is heated slightly with the small heating-loop until the rim is smooth. This will slightly thicken the wall of the very tip and make it stronger. The tip opening is ca 4-6 μm. Subsequently, the tip is sealed with a silicon membrane. This is done under a microscope, a silicon drop in the tip of a Pasteur pipette is touched by the tip of the casing. Silicon is pulled in the tip by capillary force to a thickness of ca 10 μm and subsequently cured overnight.

1.3 Guard

A guard cathode is made by etching a silver wire in saturated KCN. The etched wire is rinsed with water, HCl, water and ethanol and placed in a thin glass capillary with the tip protruding ca 3 cm. Then the glass is sealed to the silver by heating in the big heating loop.

1.4 Reference

A reference anode is made from a 100 μm thick silver wire that is chlorinated in 0.1 M HCl, by applying a potential of 2 V for 10 seconds.

1.5 Assembly

All components of the electrode are now finished and the sensor can be assembled. The components should be stored dust-free. First the casing should be fixed under the microscope. Fixing can be conveniently done with synthetic modeling clay or a clamp. Then the cathode is inserted and pushed with a micromanipulator towards the tip of the casing until the gold tip is positioned at a distance of ca 50 μm. It must be fixed with two

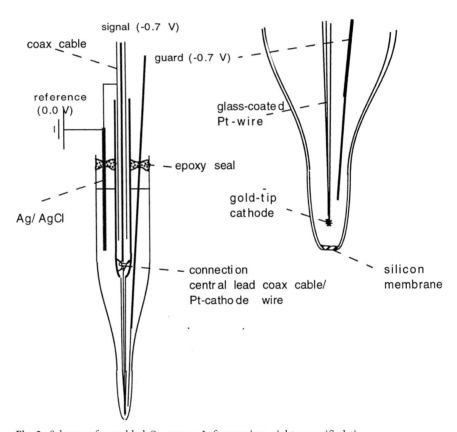

Fig. 2. Scheme of asembled O$_2$ sensor. Left overview, right magnified tip

component epoxy resin. Subsequently, the guard cathode is positioned to a distance of ca 300 μm from the tip. Finally, the reference is inserted and both the guard and reference are fixed with epoxy. Take good care that an opening remains for filling the sensor with electrolyte. The assembled sensor can be stored until filling with electrolyte.

1.6 Filling

The filling with electrolyte can be done with a 1 ml syringe from which the end is pulled in a flame to a thin tube. The electrolyte consists 0.5 M KCl, 0.15 M K_2HCO_3, 0.1 M $KHCO_3$ and a 0.1 g/l thymol (to prevent bacterial growth). First add a small amount of electrolyte as far as possible to the tip. Then by applying and releasing vacuum a few times remove the air bubble and bring the electrolyte further into the tip. If a small air bubble persists, the sensor should be placed in water that is degassed by boiling. The remaining air will diffuse through the silicon membrane in the tip into the water. This is a slow process that can take hours. Then the sensor must be filled up with electrolyte almost to the top and sealed with epoxy resin. A finished sensor is depicted in Fig 2.

Above described procedure will result in a sensor with a response time of ca 1 second. For photosynthesis measurements faster sensors are needed with a response time of 0.2 seconds. The manufacturing is slightly different: 1) the tip opening of the casing must be ca 2 μm (instead of 4-6 μm). 2) the distance between the gold cathode and the silicon membrane is ca 10 μm (instead of 50-100 μm). The fast sensors have a life time of 4-6 weeks, the normal sensors can be used for up to a year.

1.7 Connection and Calibration

The leads of the sensor must be connected to a power source and a picoamperemeter. The potential on the gold cathode and the guard is -0.75 (±0.05) V relative to the reference electrode. At this potential O_2 will be reduced:
$$0.5\ O_2 + 2e^- + H^+ \rightarrow OH^-$$

The guard cathode has a large surface and will consume all O_2 in the electrolyte. Only O_2 diffusing through the silicon membrane at the tip of the sensor can reach the gold cathode. Consequently, the current through the gold cathode is linearly proportional with the O_2 concentration near the tip of the sensor. The current between the gold cathode and the reference is measured with the picoamperemeter.

When connecting the sensor to the power source and the pico-ampere-meter care must be taken with respect to the sequence: first the reference, then the guard and then the gold cathode. If the guard or the gold cathode are connected before the reference gas bubbles may be formed in the tip, that cannot easily be removed at this stage.

After connection it takes between 10 minutes to a few hours for the sensors to stabilize. If stabilization does not occur overnight the sensors can be discarded. Since the signal is linearly proportional with the O_2 concentration a two point calibration is sufficient, mostly in air and N_2 sparged water. Thanks to the guard cathode, the offset (signal in anoxic water) is close to zero, the signal in aerated water (pA_{air}) is 60-150 pA.

Subprotocol 2
Measurements

Procedure

After calibration the sensor is ready for measurements. These are either series of concentration measurements in space, to record concentration profiles, or series of concentration measurements in time at a fixed position, to record transient concentration changes after a step change of an environmental parameter. Both type of experiments can give valuable information as will be explained hereafter.

The sensor is mounted on a micromanipulator, during which care must be taken that the leads do not disconnect. It is best to use a motorized micromanipulator, as it is much more precise than manual manipulation, and causes less disturbance. Moreover, it allows automatic profiling by a computer connected to the motor controller and the picoamperemeter. The waterphase overlaying the sample is best grounded by a reference electrode, to avoid build-up of static electricity that might destroy the sensor. Static electricity may develop when pumps for recirculation of the water phase are used or when the water phase is sparged for aeration. The microsensor is positioned above the ICS and moved to the surface, under optical guidance. The surface is a logical reference point. For profiling, the sensor the signal (pA_m) is then recorded at different depths. A complete profile comprises measurements in the boundary layer and inside the ICS. After each step one should wait for a stable signal, which takes usually 5-10 seconds. For transient measurements the sensor is positioned at any desired depth in the ICS. The calibration curve has the form of $[O_2]= (pA_m$-

offset)/(pA$_{air}$-offset) * C$_{air}$. C$_{air}$ is dependent on salinity, temperature and pressure. Reliable approximations for temperatures between 0-40 °C and salinities between 0-40‰ at atmospheric pressure can be found in the literature (Garcia and Gordon, 1992).

2.1 Determination of Local Activities from Steady State Profiles

Due to the combined effect of microbial conversion and mass transfer resistance substrate and product profiles develop inside ICS. If the ICS is impermeable for flow, diffusion is the only transport mechanism inside the matrix. Diffusional transport is driven by the concentration differences as expressed in Fick's law:

$$J = D_{eff} \frac{dc}{dx} \tag{1}$$

where J is the flux (mol m^{-2}s^{-1}), D$_{eff}$ is the effective diffusion coefficient, i.e. the D in the ICS (m^{2}s^{-1}), dc is the change in concentration (mol m^{-3}) over the distance dx (m); dc/dx is the concentration gradient. In steady state local conversion rates equal local transport rates. Assuming a constant D$_{eff}$, the mass balance becomes for a planar geometry:

$$D_{eff} \frac{d^2c}{dx^2} r \tag{2}$$

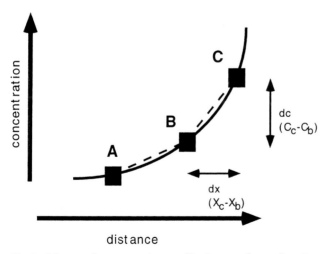

Fig. 3. Scheme of concentration profile. See text for explanation

where r is the conversion rate. With Eq. 2 the concentration profiles can be calculated for zero- and first-order kinetics, with homogeneous activity distribution. In reality, the assumptions are often too simple: conversions are of mixed order kinetics (Monod) and the distribution of activity varies in depth can be calculated.

A more direct and simple approach is to measure the relevant profiles with microsensors and derive from these the local activities. From a simple profile with 3 points as in Fig. 3, the flux between a and b and b and c can be calculated from Eq. 1:

$$J_{ab} = D_{ab} \frac{(c_a - c_b)}{(x_a - x_b)} \text{ and } J_{bc} = D_{be} \frac{(c_b - c_c)}{(x_b - x_c)} \tag{3}$$

This gives the best estimates for the fluxes through the intermediate points between the measurements, $0.5(x_a+x_b)$ and $0.5(x_b+x_c)$. If the system is in steady state, then the difference in fluxes through these points is equal to the local conversion in point b, which is approximated by:

$$r_b = \frac{J_{ab} - J_{ab}}{0.5(x_a + x_b) - 0.5(x_b + x_c)} \tag{4}$$

If D_{eff} is constant with depth and we measured with constant depth intervals Δx then

$$r_b = D_{eff} \frac{c_a - 2c_b + c_c}{\Delta x^2} \tag{5}$$

This approach needs high spatial resolution microprofile measurements, so one should measure as many points as possible. However, the step size should not be smaller than 2x the diameter of the sensor tip. Since noise is magnified, noisy profiles give poor results. Such data can be used, but a smoothing procedure is recommended, e.g. by averaging each point with its neighbor points. Alternatively, the profiles can be fitted with a high-order polynomium. Then the local fluxes are given by the product of the derivative of that polynomial function and D_{eff}. The local activities are calculated by the product of the second derivative of the function and D_{eff}. When D_{eff} varies in depth, the local fluxes must be calculated with the local D_{eff}, using Eq. 3, and local activities using Eq. 4. All these calculations can be conveniently done with a spreadsheet.

The diffusion coefficient is needed to calculate the fluxes from the microprofiles. Unfortunately, its value is often uncertain. For all compounds measured by microsensors the molecular diffusion coefficient (D_w) is rather well known, but of more interest is the diffusion coefficient inside the ICS, D_{eff}. D_{eff} is always lower than the D_w, due to the reduced surface available for diffusion (porosity) and the increased path length caused by obstacles (tortuosity). D_{eff}/D_w is typically 0.8-1 in biofilms (Christensen and Characklis, 1990), close to 1 in flocs (Christensen and Characklis, 1990), 0.7-0.9 in microbial mats (Canfield and Des Marais, 1993) and 0.3-1 in sediments (Ullman and Aller, 1982).

2.1.1 Example

Microbial mats are communities of photosynthetic algae, mainly *Cyanobacteria*, and respiratory bacteria. These are almost closed systems in which most of the photosynthetically bound organic matter is reoxidized, both by aerobic respiration and by sulfate reduction(Jørgensen and Cohen, 1977). Although the photosynthetic rates are among the highest measured (Cohen et al., 1977; Revsbech, 1994), the net accretion of these mats is negligible. Microbial mats are typical for extreme environments (high salinity or temperature) and characterized by the absence of bioturbation, as insects and worms cannot survive the environment. They are rather thick and smooth, therefore an ideal object for microsensor analysis. Although the ecological relevance of these rather rare systems is limited, they form interesting model systems for microbial ecologists. Fig. 4 depicts steady state O_2 profiles in light and dark incubation. In the light, oxygen penetrates down to 2 mm deep. Oxygen is produced in the top zone and diffuses both upwards into the water phase and downwards. The middle panel of Fig. 4 shows the local fluxes, according to Eq. (3). The steepest gradients are translated to the highest local fluxes. The areas of the highest conversion rates are those where fluxes are changing rapidly, as can be seen in the right panel of Fig. 4, which is calculated with Eq. (4) and smoothed. Negative values indicate O_2 production (negative consumption) thus photosynthesis. The thick lines give the average activities over a certain area. Clearly, in the very top layer of ca 250 μm, no photosynthesis takes place. This is possibly an adaptation of the algae to excess light, that causes them to migrate downwards. In the dark, oxygen penetrates less than 400μm into the mat, and only respiration is observed. Also in darkness the very top region of the mat (150 μm) does not show much activity, possibly because no photosynthates are formed that can fuel the microbial conversions.

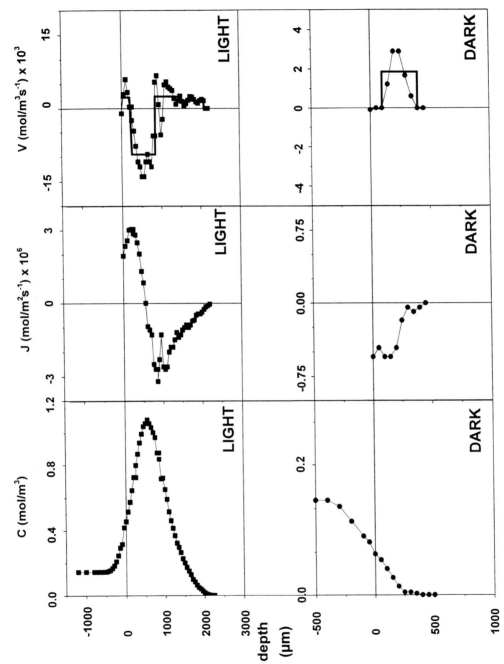

Fig. 4. Oxygen concentration profiles (left panels), fluxes (middle panels) and activities (right panels) in a microbial mat from Solar Lake (Sinai), incubated in light (top panels) and darkness (bottom panels). The thick solid lines indicate the average microbial activity over an area. Note the differences in scaling of the dark and light profiles

2.2 Local Activities from Transients

In steady state the sum of consumption (R), production (P) and transport in each point in the matrix equals zero:

$$dC(x)/dt = D_{eff}d^2C(x)/dx^2 + P(x) - R(x) = 0 \qquad (6)$$

$$\text{or}: \ P(x) = -(D_{eff}d^2C(x)/dx^2 - R(x)) \qquad (7)$$

If the production is stopped instantaneously, then

$$dC(x)/dt = -D_{eff}d^2C(x)/dx^2 - R(x)) \qquad (8)$$

$$\text{and}: \ P(x) = dC(x)/dt \qquad (9)$$

Consequently, directly after microbial conversion is stopped, the initial change in product or substrate concentration is equal to the microbial conversion rate before the process was stopped.

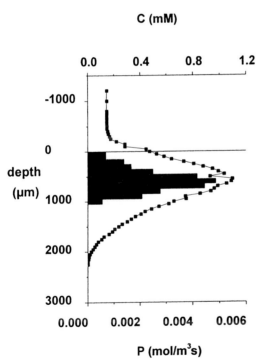

Fig. 5. Photosynthesis profile recorded with the fast light-dark shift (gray area), and concentration profile

The best example of this approach is the determination of photosynthetic activities using O_2 microsensors using the light-dark method (Revsbech and Jørgensen, 1983), as photosynthesis can be stopped instantaneously by darkening. For accurate determination, fast darkening is needed (within a few milliseconds) and fast sensors are needed (response times 2 concentration decreases, at a rate that decreases within 2 seconds. The initial rate of the decrease equals the photosynthesis before darkening. By measuring the photosynthesis rate at different depths of a photosynthetic mat a photosynthesis profile can be recorded. Of course, after each measurement the steady state must be restored by illumination until a constant O_2 concentration is reached.

References

Ammann, D., Lanter, F., Steiner, R. A., Schulthess, P., Shio, Y. and Simon, W. (1981) Neutral carrier based hydrogen-ion selective microsensor for extra- and intracellular studies. Anal. Chem. 53:2267-2269

Canfield, D. and Des Marais, D. J. (1993) Biochemical cycles of carbon, sulfur, and free oxygen in a microbial mat Geochimica Cosmochimica Acta 57:3971-3984

Christensen, B. E. and Characklis, W. G. (1990) Physical and chemical properties of biofilms. In: W. G. Characklis and K. C. Marshall (eds) Biofilms. Wiley and sons, pp 93-130

Cohen, Y., Krumbein, W. E. and Shilo, M. (1977) Solar Lake (Sinai) 2 Distribution of photosynthetic microorganisms and primary production. Limnol.Oceanogr. 22:609-619

Damgaard, L. R. and Revsbech, N. P. (1997) A microscale biosensor for methane. Anal. Chem. 69:2262-2267

De Beer, D. (1990) . University of Amsterdam, Amsterdam.

De Beer, D., Glud, A., Epping, E. and Kühl, M. (1997a) A fast responding CO_2 microelectrode for profiling sediments, microbial mats and biofilms. Limnology Oceanography 42:1590-1600

De Beer, D., Schramm, A., C.M., S. and Kuhl, M. (1997b) A nitrite microsensor for profiling environmental biofilms. Applied Environmental Microbiology 63:973-977

De Beer, D. and Sweerts, J. P. R. A. (1989) Measurements of nitrate gradients with an ion-selective microelectrode. Analytica Chimica Acta 219:351-356

De Beer, D., Sweerts, J.-P. R. A. and van den Heuvel, J. C. (1991) Microelectrode measurement of ammonium profiles in freshwater sediments FEMS Microbiology Ecology 86:1-6

De Beer, D. and Van den Heuvel, J. C. (1988) Response of ammonium-selective microelectrodes based on the neutral carrier nonactin. Talanta 35:728-730

De Beer, D., van den Heuvel, J. C. and Ottengraf, S. P. P. (1993) Microelectrode measurements of the activity distribution in nitrifying bacterial aggregates. Applied Environmental Microbiology 59:573-579

De Boer, J. P., Cronenberg, C. C. H., De Beer, D., Van Den Heuvel, J. C., De Mattos, M. J. T. and Neijssel, O. M. (1993) pH and glucose profiles in aggregates of bacillus-laevolacticus. Appl. Environ. Microbiol. 59:2474-2478

Garcia, H. E. and Gordon, L. I. (1992) Oxygen solubility in seawater: better fitting equations Limnol. Oceanogr. 37:1307-1312

Hartley, A. M., House, W. A., Leadbeater, B. S. C. and Callow, M. E. (1996) The use of microelectrodes to study the precipitation of calcite upon algal biofilms. J. Coll. Interfac. Sci. 183:498-505

Hinke, J. (1969) Glass microelectrodes for the study of binding and compartmentalisation of intracellular ions. In: M. Lavallee, O. F. Schanne and N. C. Herbert (eds) Glass microelectrodes. Wiley, pp 349-375

Jensen, K., Revsbech, N. P. and Nielsen, L. P. (1993) Microscale distribution of nitrification activity in sediment determined with a shielded microsensor for nitrate. Appl. Environm. Microbiol. 59:3287-3296

Jensen, K., Sloth, N. P., Risgaard-Petersen, N., Rysgaard, S. and Revsbech, N. P. (1994) Estimation of nitrification and denitrification from microprofiles of oxygen and nitrate in model sediment systems. Appl. Environm. Microbiol. 60:2064-2100

Jeroschewski, P., Steukart, C. and Kühl, M. (1996) An amperometric microsensor for the determination of H_2S in aquatic environments. Anal.Chem. 68:4351-4357

Jørgensen, B. B. and Cohen, Y. (1977) Solar Lake (Sinai) 5 The sulfur cycle of the bentic cyanobacterial mats 657-666

Kühl, M. and Jørgensen, B. B. (1992) Microsensor measurements of sulfate reduction and sulfide oxidation in compact microbial communities of aerobic biofilms. Appl. Environm. Microbiol. 58:1164-1174

Kühl, M., Steuchart, C., Eickert, G. and Jeroschewski, P. (1998) A H_2S microsensor for profiling biofilms and sediments: Application in an acidic lake sediment. Aquat. Microb. Ecol. 15:201-209

Larsen, L. H., Kjaer, T. and Revsbech, N. P. (1997) A microscale NO_3^- biosensor for environmental applications. Anal. Chem. 69:3527-3531

Larsen, L. H., Revsbech, N. and Binnerup, S. J. (1996) A microsensor for nitrate based on immobilized denitrifying bacteria. Appl. Environm. Microbiol. 62:148-1251

Lens, P., de Beer, D., Cronenberg, C., Ottengraf, S. P. P. and Verstraete, W. (1995) The use of microsensors to determine population distributions in UASB aggregates. Wat. Sci. Technol. 31:273-280

Lens, P., de Beer, D., Cronenberg, C. C. H., Houwen, F. P., Ottengraf, S. P. P. and Verstraete, W. (1994) Heterogeneous distribution of microbial activity in methanogenic aggregates: pH and glucose microprofiles. Appl. Environm. Microbiol. 59:3803-3815

McConnaughey, T., D. and Falk, R. H. (1991) Calcium proton exchange during algal calcification. Biol. Bull. 180:185-195

Revsbech, N. P. (1989) An oxygen microelectrode with a guard cathode. Limnol. Oceanogr. 55:1907-1910

Revsbech, N. P. (1994) Analysis of microbial mats by use of electrochemical microsensors: recent advances. In: L. Stal and P. Caumette (eds) Microbial mats. Springer Verlag, pp 135-147

Revsbech, N. P. and Jørgensen, B. B. (1983) Photosynthesis of benthic microflora measured with high spatial resolution by the oxygen microprofile method: Capabilities and limitations of the method. Limnol. Oceanogr. 28:749-756

Revsbech, N. P. and Jørgensen, B. B. (1986) Microelectrodes: their use in microbial ecology. Adv.Microbial Ecol. 9:293-352

Revsbech, N. P., Jørgensen, B. B., Blackburn, T. H. and Cohen, Y. (1983) Microelectrode studies of the photosynthesis and O2, H2S and pH profiles of a microbial mat. Limnol. Oceanogr. 28:1062-1074

Revsbech, N. P., Jørgensen, B. B. and Brix, O. (1981) Primary production of microalgae in sediments measured by oxygen microprofile, $H^{14}CO_3^-$ fixation, and oxygen exchange methods Limnol.Oceanogr. 26:717-730

Revsbech, N. P., Madsen, B. and Jørgensen, B. B. (1986) Oxygen production and consumption in sediments determined at high spatial resolution by computer simulation of oxygen microelectrode data Limnology and Oceanography 31:293-304

Revsbech, N. P., Nielsen, L. P., Christensen, P. B. and Sorensen, J. (1988) A combined oxygen and nitrous oxide microsensor for denitrification studies. Appl. Environ. Microbiol. 45:2245-2249

Revsbech, N. P. and Ward, D. M. (1983) Oxygen microelectrode that is insensitive to medium chemical composition: Use in an acid microbial mat dominated by *Cyanidium caldarum*. Appl. Environ. Microbiol. 45:755-759

Sweerts, J.-P. R. A., Bar-Gillisen, M. J., Cornelise, A. A. and Cappenberg, T. E. (1991) Oxygen consuming processes at the profundal and littoral sediment-water interface of a small meso-eutrophic lake (lake Vechten, the Netherlands) Limnology and Oceanography 36:1124-1133

Sweerts, J.-P. R. A. and de Beer, D. (1989) Microelectrode measurements of nitrate gradients in the littoral and profundal sediments of a meso-eutrophic lake (lake Vechten, The Netherlands). Applied Environmental Microbiology 55:754-757

Ullman, W. J. and Aller, R. C. (1982) Diffusion coefficients in nearshore marine sediments. Limnol. Oceanogr. 27:552-556

Wijffels, R. H., Eekhof, M. R., de Beer, D., van den Heuvel, J. C. and Tramper, J. (1995) Pseudo-steady state oxygen-concentration profiles in an agar slab containing growing *Nitrobacter agilis*. J. Fermentation & Biotech 79:167-170

Biomass Gradients

RENÉ H. WIJFFELS

Introduction

Micro-organisms which are entrapped in gel beads with retention of their viability can subsequently be cultivated. The immobilized cells will thus grow within the support material. They do that by cell division and as a result micro-colonies are formed. Initially the biomass concentration is low. The small colonies that are formed have the same size all over the bead. As the biomass concentration increases, however, effects of diffusion limitation may become important, resulting in non-homogeneous growth of biomass across the beads: near the bead surface larger colonies are formed than in the core of the bead (Gosmann and Rehm 1986, Khang et al. 1988, Wada et al. 1980, Chibata et al. 1983, Wijffels and Tramper 1989). Eventually, the colonies may expand in such a way that they confluence and form a dense internal biofilm (Monbouquette et al. 1990).

Total biomass concentration in the support material is relatively easy to determine as has been described in Chapter 7. Local concentrations are more complicated to determine. Shells can be dissolved from gel particles in which cell concentrations can be determined by cell counts or turbidity measurements (Boross et al. 1989, Lefebvre and Vincent 1992, Walsh et al. 1993). A drawback of this method is that usually shells of a thickness larger than 50µm are obtained. As profiles tend to be very steep, a higher resolution than this is required.

More accurate methods have been developed for artificially immobilized cells. Ultra-thin sections from the centre of the beads are sliced and within these sections the volumetric fraction of colonies can be deter-

René H. Wijffels, WageningenUniversity, Food and Bioprocess Engineering Group, P.O. Box 8129, Wageningen, 6700 EV, The Netherlands (*phone* +31-0317-484372; *fax* +31-0317-482237; *e-mail* rene.wijffels@algemeen.pk.wag-ur.nl)

mined (Karel and Robertson 1989, Kuhn et al. 1991, Monbouquette et al. 1990, Wijffels et al. 1991, 1995, Hunik et al. 1993, 1994, Leenen et al.1997).

Profiles can be quantified by image analysis. The coordinates of the centres and the radii of the colonies are determined in relation to the centre or the surface of the bead (Wijffels et al. 1991). For this analysis, the slices have to be very thin to prevent colonies from overlapping. Samples for image analysis are usually stained:

1. To increase contrast. General staining methods with e.g. toluidine blue are used (Wijffels et al. 1991, 1995).

2. To distinguish growing cells from non growing cells. Cell components or reactions which are related to cell growth are stained:
   - RNA (Monbouquette et al. 1990)
   - DNA synthesis (Kuhn et al. 1991)
   - Addition of the yellow coloured dimethylthiazol salt which is transferred to MTT formazan which has a blue colour (Al-Rubeai and Spier 1989)
   - ^{35}S (Karel and Robertson 1989) Immobilized cells were stained with a b-emitting isotope ^{35}S which was supplied in the form of sulphate and is incorporated in proteins during synthesis. Consequently, growth can be monitored as ^{35}S activity. Biomass profiles were observed with a light microscope after exposure of the slices to an autoradiographic emulsion and development.
   - Fluorescein-diacetate/lissamine green staining (Leenen et al. (1996). Fluorescein-diacetate labelling was used to label viable cells, and cells with degenerated membranes were labelled with lissamine green.

3. To specify certain strains. Hunik et al. (1993, 1994) used a labelling technique with fluorescent antibodies to distinguish two species of immobilized cells. Two ultra-thin centre sections were labelled with antibodies. One with a rabbit antibody for *Nitrosomonas* and the other with a rabbit antibody for *Nitrobacter*. The rabbit antibodies were labelled with a goat antibody which was conjugated with fluorescein isothiocyanate (FITC). As a result fluorescent colonies were obtained.
   Although the slicing methods used in combination with image analysis are more accurate than scraping or dissolving of shells, there are also some disadvantages. Samples have to be cut in thin slices (< 5 µm) to prevent overlap of the colonies. To be able to make such thin sections with a microtome the samples have to be embedded in a resin and usually chemical fixation is necessary. Shrinkage of the samples has

been observed during preparation (van Neerven et al. 1990) and this may affect the results obtained. More direct methods would improve the quality of the results.

4. Non-invasive methods. Such direct techniques have been presented by Worden and Berry (1992) and Hüsken et al. (1996). They mounted a thin slab of gel containing immobilized cells on a microscope slide. One side of the slab was in contact with nutrient medium and growth of cells could be observed through a microscope. Apart from being a direct technique, this is non-invasive as well, which means that it is possible to study the dynamics of the process.

Outline

The methods given in this chapter describe quantification of the profiles by analyzing micrographs of median sections of carrageenan-gel beads by image analyses. Immobilized cells profiles are quantified by determination of the colony size as a function of bead radius. Three methods will be distinguished. The first method describes the general technique used for sectioning and analysis. No distinction is made on cell types. In a second method biomass is labelled to discriminate between viable and not viable cells. In the third method cells are labelled with antibodies specifically binding to certain organisms. With the last method a discrimination between the different strains can be made. The second and third method will concentrate more on labelling techniques as sectioning of the beads and image analysis is comparable to the first method. The last method is not based on sectioning. A gel slab can be observed microscopically when it is mounted in a microscope reactor.

Materials

Non-specific biomass gradients

- mould for a cylinder of 3 mm diameter
- LKB 8800 Ultramicrotome III with a glass knife
- light microscope

- 0.1 M Na-cacodylate, pH 7.5

- KCl

- κ-carrageenan

- 2.5 % (w/v) glutaraldehyde

- 1.0 % OsO_4 (w/v)

- ethanol

- London Resin White (Bio-rad)

- image analyser (e.g. Quantimet 970, Cambridge Instruments or Magis-can image analysis system with GENeral Image Analysis Software). Important is that the size of one pixel is not larger than the size of a single cell.

Distinguish growing cells from non-growing cells

- fluorescein-diacetate (FDA)

- lissamine green (Gurr)

- razor blade

- fluorescence microscope (Nikon, Labophot)

- UV-light (dichroic mirror DM 400, excitation filter Ex 365 and barrier filter BA 420)

Specification of strains

- KCl

- K_2HPO_4

- KH_2PO_4

- NH_4Cl

- $NaBH_4$

- ethanol

- paraformaldehyde

- PEG 1500 (Merck 807489)

- PEG 4000 (Merck 807490)

- PEG 6000 (Merck 807491)

- bovine serum albumin (BSA) fraction V (Merck 12018)

- rabbit-anti-*N. europaea* (Verhagen and Laanbroek 1991)

- rabbit-anti-*N. agilis* (Laanbroek and Gerards 1991)

- goat-anti-rabbit FITC conjugate (Sigma F6005)

- Citifluor in glycerol solution (AF2, van Loenen Instruments, The Netherlands)

- mould for a cylinder of 3 mm diameter

- Leitz rotary microtome equipped with a Kulzer knife

- poly-L-lysine coated glass slide

- Microphot Nikon FXA microscope with a fluor pan Nikon objective (10x/0.5 NA)

- DM500 dichroic mirror

- B470-490 excitation filter

- BA 520-560 emission filter

Non-invasive techniques

- microscope reactor

- microscope

- gear pump

Procedure

Non-specific biomass gradients

Sectioning and staining

In order to obtain thin sections the beads were dehydrated and embedded in a resin. Chemical fixation prior to dehydration appeared to be necessary. The procedure was based on the work of Van Neerven et al. (1990).

1. Wash the beads twice for 5 minutes in buffer (0.1 M Na-cacodylate, pH 7.5). To keep the carrageenan beads rigid 0.1 mmol · l^{-1} KCl needs to be added to the Na-cacodylate during all washing and fixation.

2. Fixate the beads for 1.5 hours in the same buffer containing 2.5 % (w/v) glutaraldehyde.

3. Wash the beads 3 times for 10 minutes in the Na-cacodylate buffer.

4. Fixate the beads for 1.5 hours in the buffer containing 1.0 % OsO$_4$ (w/v).

5. Wash the beads 3 times for 10 minutes in the Na-cacodylate buffer.

6. Replace the buffer with a pure ethanol solution in a series of gradually increasing ethanol concentrations: 10, 30, 50, 70, 90 and twice 100 %. A decreasing KCl gradient from 0.09 to 0.03 mmol · l^{-1} needs to be used in the increasing ethanol series of 10, 30, 50 and 70 %. No KCl needs to be added in the two ultimate dehydration steps. Each step is allowed to equilibrate for 10 minutes.

7. Embed the dehydrated beads in London Resin White (Bio-rad). Therefore, the beads are placed in ethanol-London Resin White solutions in the ratios 5:1, 3:1, 1:1 and 1:3, respectively, before treatment with pure London Resin White. Each step is allowed to equilibrate for 30 minutes. Treatment with pure London Resin White is executed in two steps. The first step for 60 minutes and the second for 24 hours. Polymerization of the resin takes place in a gelatine mould at 60 °C for 24 hours.

8. Section the mould containing the beads on a LKB 8800 Ultramicrotome III with a glass knife.

9. Stain the sections with toluidine blue to improve the contrast between the gel and the colonies.

Image analysis

1. Observe the samples at a magnitude of 40 on a light microscope and take photographs using a Kodak Technical Pan film 2415 at 50 ASA. The prints (final magnification 150x) are used for image analysis with a

Quantimet 970 (Cambridge Instruments). It is also possible to analyze the microscopic image directly with image analysis.

2. By image analysis the coordinates of the centres of the colonies are determined in relation to the centre of the bead. Also the radii of the colonies are determined. For determination of the volumetric fraction of colonies the beads are divided into fictive shells by test lines. In those shells the areal density is determined according to the principle of Rosiwal, which is that the areal density is equal to the fractional length of a test line that intersects the objects (Weibel 1979, 1980). In this case the test line is a circle with a fixed distance to the centre of the beads (Figure 1). The fraction of the test line that sections the colonies can be calculated from the distance from the centre of the bead to the centre of the colony minus the distance from the centre of the bead to the drawn test line and the colony radius. In the core section of the beads the areal density was directly determined in three area classes, because the distribution of the colony fraction is relatively random there. In thin slices the areal density is equal to the volumetric density according to the principle of Delesse (Weibel 1979, 1980).

$$c = \frac{\pi}{\pi + 2L/r_c} \qquad (1)$$

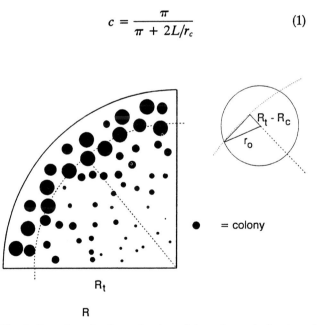

Fig. 1. Procedure for determination of the volumetric fraction of biomass; r_o = observed colony radius, R_c = distance between centre of colony and bead, R = bead radius and R_t = radius test line

As the observed areas are projections of the observed colonies situated in slices with a thickness L, it may be necessary to correct for over-estimations as more images are projected for thicker slices (the Holmes effect) and under-estimations as images can overlap (Weibel 1979, 1980, Hennig 1969, Underwood 1972).

As the slices are thin compared to the colony radii, overlap may be neglected (Hennig 1969). For the Holmes effect, the observed radii can be corrected by a factor defined as (Hennig 1969):

$$c = \frac{\pi}{\pi + 2\frac{L}{r_c}} \tag{1}$$

As the observed colony radii are expectations of the colony radii, the colony radii can be obtained from:

$$E(\underline{D}) = 2\, r_c \frac{1 + \frac{\pi}{2}\frac{r_c}{L}}{1 + 2\frac{r_c}{L}} \tag{2}$$

and the observed projections can be corrected individually.

The overall biomass concentration needs to be determined separately, e.g. by a protein assay as has been described in Chapter 7.

Distinguish growing cells from non-growing cells

Two staining techniques were used to discriminate between viable cells and cells with a degenerated cell membrane (here designated as dead). The fluorescein-diacetate labelling was used to label viable cells, and cells with degenerated membranes were labelled with lissamine green.

The fluorescein-diacetate and lissamine-green staining techniques were applied simultaneously to one sample. By changing from bright-field mode to UV the viable or dead cells could be counted in the same image.

Fluorescein diacetate (FDA)   The fluorescein-diacetate (FDA) labelling is based on the activity of esterases present in metabolic-active cells, which catalyse the reaction:
fluorescein diacetate (FDA)® fluorescein + 2 acetate

The colourless FDA is transported through an intact cell membrane. Inside viable cells esterases convert FDA to fluorescein. Fluorescein cannot leave the cell through an intact cell membrane and is fluorescent under UV-excitation. The fluorescein will thus accumulate inside viable cells. Dead cells do not contain esterases and remain colourless (Chrzanowski et al., 1984; Heslop-Harrison and Heslop-Harrison, 1970).

The procedure given below was selected after testing incubation times of 1, 5 or 30 minutes and addition of 0.2, 0.5 or 1.0 ml FDA-solution. For each strain or support material it will be necessary to test the efficiency of the method.

1. A stock-solution of 0.5% FDA (Sigma) in acetone needs to be stored at -20°C. Just before use, a 25 times dilution in growth medium is prepared.

2. Stain immobilized cells for 5 minutes by adding 0.5 ml FDA-solution to 10 gel beads.

3. Cut the gel beads with a razor blade.

4. Observe the sections with a fluorescence microscope (Nikon, Labophot) under UV-light (dichroic mirror DM 400, excitation filter Ex 365 and barrier filter BA 420).

Lissamine green is a selective stain for the cytoplasm of degenerating and degenerated cells, because it only permeates disrupted cell membranes. The lissamine green will accumulate as a result of the irreversible binding. It can not enter viable cells, thus no accumulation will occur there. Dead cells treated with lissamine green will, therefore, appear green under a microscope in bright-field mode.

Lissamine green

The procedure given below was selected after testing various incubation times (1,5 or 30 minutes) and lissamine-green quantities (0.2, 0.5 or 1.0 ml stock-solutions). For each strain or support material it will be necessary to test the efficiency of the method.

1. A stock-solution of 0.2% lissamine green (Gurr) in medium needs to be stored at -20°C.

2. Stain immobilized cells for 5 minutes by adding 0.5 ml lissamine-green solution to 10 gel beads.

3. Cut the gel with a razor blade.

4. Observe the sections with a microscope (bright field). Viable cells will be colourless and dead cells will be stained green.

Specification of strains

Sectioning was similar to the procedure as described in the 'paragraph non-specific biomass gradients'. Nevertheless, there were small differences in the method used. Therefore, the sectioning method used is given again but

Sectioning and staining

can be replaced by the method described previously. For all methods used you should be aware that auto fluorescence can seriously disturb the final observations. To diminish auto fluorescence, variations in sectioning and fixation should be tested. All steps described below need to be done at room temperature and high air humidity (above 60%).

The protocol given below was done with antibodies raised against *Nitrosomonas europaea* and *Nitrobacter agilis*. This protocol can be applied for labelling for any immobilized strain with their specific antibodies.

1. Harvest gel beads and store in a 0.1 M KCl solution for one hour.

2. Wash gel beads twice with a potassium phosphate buffer (PPB, 8.7 g of K_2HPO_4, 6.8 g of KH_2PO_4, 7.45 g of KCl per dm^3 and a pH of 7.4) for 10 minutes.

3. Fixate gel beads in paraformaldehyde (3% w/v in PPB) for two hours.

4. Remove paraformaldehyde by washing with PBB three times for 10 minutes.

5. Wash gel beads twice with 0.1 M KCl for 10 minutes.

6. Dehydrate beads in ethanol-KCl solutions starting with 10% ethanol-0.09 M KCl and successively followed by 20% ethanol- 0.08 M KCl, 30% ethanol- 0.07 M KCl, 50% ethanol- 0.05 M KCl, 70% ethanol- 0.03 M KCl, 90% ethanol, 100% ethanol and finally ending with an additional 100% ethanol step. The gel beads are kept 20 minutes in each solution.

7. Replace 100% ethanol by Poly Ethylene Glycol (PEG) with an infiltration range at 55 °C. PEG is a 1:2 mixture of PEG 4000 (Merck 807490) and PEG 1500 (Merck 807489). Change the PEG/ethanol ratio successively from 1:10 to 1:4, 1:1, 4:1 and 10:1. The gel beads are kept for 30 minutes in each solution. Finally, keep the beads in 100% PEG for 60 minutes.

8. Polymerize the PEG infiltrated gel beads in a mould by cooling to room temperature.

9. Section the mould containing the beads with a Leitz rotary microtome equipped with a Kulzer knife. Thickness of the sections is 2 μm.

10. Collect sections in a droplet of (40% w/v) PEG 6000 (Merck 807491) solution in phosphate buffered saline (PBS, pH 7.4) and attach to a poly-L-lysine coated glass slide. For this, wash a microscopic glass slide in 96% ethanol for 30 min. and coat the glass slides with 0.1 % (w/v) poly-lysine solution.

11. Remove PEG by rinsing with PBS for 10 min.

12. Block aldehyde with NH_4Cl (0.1M) for 5 min.

13. Block aldehyde with $NaBH_4$ (0.5 mg/dm^3) for 5 min.

14. Wash with PBS for 10 min.

15. Block background with 1% (w/v) bovine serum albumin (BSA) fraction V (Merck 12018) in PBS for 60 min.

16. Wash with 0.1 % (w/v) BSA in PBS, twice for 30 min each.

17. Label with rabbit-anti-*N. europaea* (Verhagen and Laanbroek 1991) or rabbit-anti-*N. agilis* (Laanbroek and Gerards 1991) (diluted 1:300 in PBS) for 45 min.

18. Wash with 0.1% (w/v) BSA in PBS, four times for 5 min each.

19. Label with goat-anti-rabbit FITC conjugate (Sigma F6005, diluted 1:40 in PBS) for 45 min. This step needs to be done in the dark.

20. Wash with PBS, 7 times for 6 min each.

21. Mount the labelled samples in a Citifluor in glycerol solution (AF2, van Loenen Instruments, The Netherlands).

Control for auto fluorescence of the colonies by omitting the labelling with the goat-anti-rabbit FITC conjugate from the staining protocol. Also test the non-specific binding of the goat-anti-rabbit-FITC. For this the labelling with rabbit-anti-*N. europaea* or rabbit-anti-*N. agilis* was omitted from the protocol. Both rabbit-anti-*Nitrosomonas europaea* and rabbit-anti-*Nitrobacter agilis* antibodies are tested with cells from the pure cultures.

1. Examine the labelled samples microscopically (Microphot Nikon FXA microscope with a fluor pan Nikon objective (10x/0.5 NA). The FITC fluorescence was observed with a DM500 dichroic mirror, B470-490 excitation filter and BA 520-560 emission filter. Photomicrographs were recorded on Kodak Ektachrome P800/1600 film (5020 EES at 800 ASA). The colony distribution was determined with a Hitachi camera unit attached to the microscope.

Microscopy and image analysis

2. For image analysis reference is made to the method described in the paragraph 'non-specific biomass gradients'.

Non-invasive techniques

A microscope reactor was developed in which growth of immobilized cells
could be observed continuously. In a perspex plate a channel and gel chamber were cut. The perspex plate was covered with a glass plate which was
tightened to the perspex plate with a metal plate as is shown in Figure 2a. A
picture of the reactor is shown in Figure 2b (Hüsken et al. 1996).

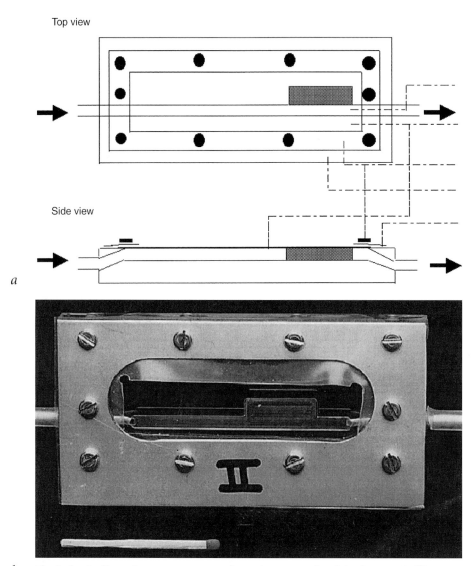

Top view

Side view

a

b   **Fig. 2a, b.** On-line microscope reactor; schematic presentation (*a*), photograph (*b*)

1. A piece of gel needs to fit exactly in the gel chamber of the reactor. It is fixed by covering it with the glass plate. Medium is pumped continuously along one side of the slab. The transparent on-line microscope reactor is placed on the table of a light microscope which can be connected to an image-analysis system. The slab (20 x 5 x 1 mm) is observed microscopically in order to study the development of the micro colonies. The diameter of the channel (d) is 1.5 mm and the length of flow (L) is 4 cm.

2. The reactor is continuously operated by supplying growth medium. Medium is pumped through the microscope reactor continuously by a gear pump. A gear pump is used to prevent pulses in flow along the slab.

3. Before entering the microscope reactor the medium needs to be saturated with air.

Results

Non-specific biomass gradients

Sections with a thickness of 3 and 4 µm, respectively, were analysed. Figure 3 shows one of the results obtained directly from image analysis. The apparent colony radii are given as a function of the radial position within the bead. These results were converted into relative biomass concentrations as shown in Figure 4 for two samples. It is shown that 90 % of the immobilized biomass is situated in an outmost shell of about 140 µm. The maximum, with a thickness of about 20 µm, was reached between a relative radius of 0.94 and 0.96. At a relative radius exceeding 0.96, the biomass concentration decreased. The decrease is caused by the fact that colonies are spherical.

Distinguish growing cells from non-growing cells

The FDA and lissamine-green staining procedures normally used for suspended cells were also used for immobilized cells. We expected that adaptations might be necessary as reagents need to diffuse to the cells. Therefore, different incubation times and amounts of FDA or lissamine-green solutions were investigated.

radius micro-colony [μm]

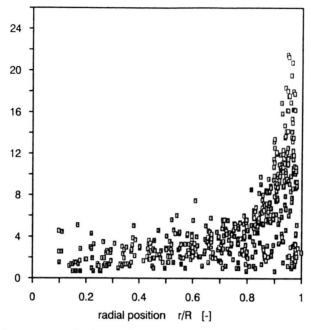

radial position   r/R   [-]

Fig. 3. Measured colony radii as a function of radial position within the beads

All combinations tested (incubation times of 1, 5 and 30 minutes and FDA or lissamine-green solutions of 0.2, 0.5 or 1.0 ml) gave approximately the same results; all cells were stained with FDA or lissamine green. Therefore, we concluded that FDA and lissamine green diffuse into the matrix easily. In the further research an incubation time of 5 minutes with 0.5 ml FDA or lissamine-green solution was used.

The viability and development of a biomass profile of immobilized *Nitrosomonas europaea* that was grown continuously for 30 days were followed microscopically during the experiment with the aid of the FDA and lissamine-green staining techniques. Typical images of sections of gel beads with immobilized *Nitrosomonas europaea* are shown in Figure 5; Figure 6 is a schematic representation of these images. Dead cells are dark-green in bright field (I) and not visible in UV (II), viable cells are colourless in bright field (I) and fluorescent in UV (II). During the first 4 days the biomass was homogeneously distributed in the gel bead and all cells were viable. From approximately day 7 on a biomass profile became evident. In an outer layer of approximately 125 μm colonies with a

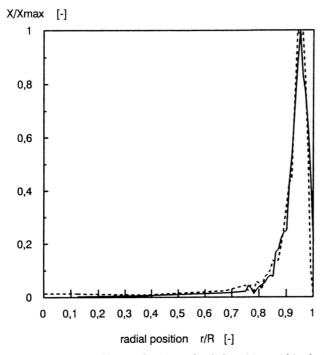

Fig. 4. Biomass profile as a function of radial position within the beads

diameter of about 11 µm developed, while in the centre of the bead the colonies were clearly smaller; all biomass was still viable. During the next days the biomass profile became steeper. After 10 days 90% of the detected biomass was still viable and large colonies with a diameter around 25 µm had developed near the surface of the bead. In these colonies both dead and viable cells were detected by showing both fluorescence in UV and a few green cells in bright field. After 13 days large colonies (diameter 30 µm) were present at the edge of the bead; in a layer of approximately 80 µm colonies with diameters larger than 10 µm were present and in deeper layers smaller colonies were observed.

Specification of strains

The volumetric biomass distribution shown in Figure 7 was based on data from gel beads at the end of the continuous growth experiment (Hunik et al. 1993).

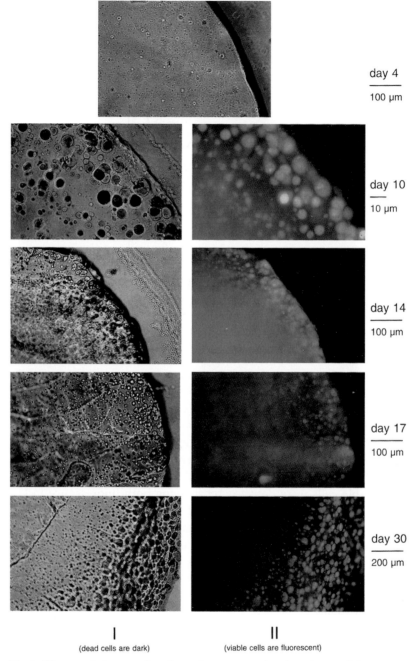

day 4

100 µm

day 10

10 µm

day 14

100 µm

day 17

100 µm

day 30

200 µm

I
(dead cells are dark)

II
(viable cells are fluorescent)

Fig. 5. Microscopic images of sections of carrageenan beads with immobilized *Nitrosomonas europaea* at day 4, 10, 14, 17 and 30. *I*) image in bright field. *II*) image in UV light

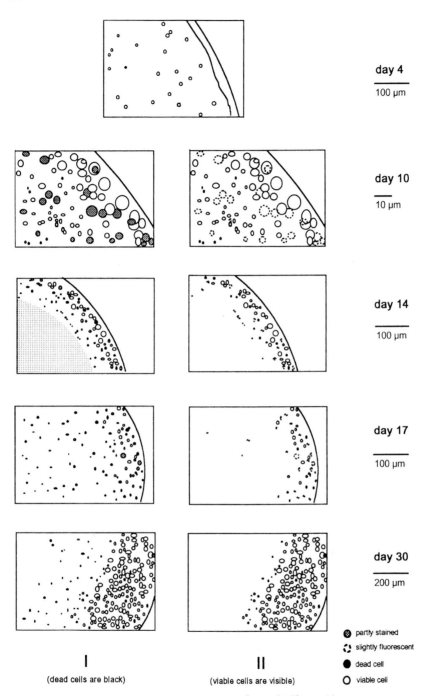

Fig. 6. Schematic representation of the images shown in Figure 4A

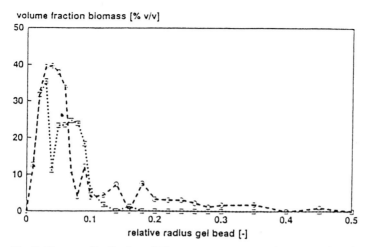

Fig. 7. Biomass distribution of *Nitrosomonas europaea* (– – I – –) and *Nitrobacter agilis* (.. n ..) after 49 days of cultivation. Volume fraction of biomass as function of the gelbead radius (0 = edge of the bead)

With the labelling technique described it was possible to measure a biomass volume distribution of two immobilized species.

Non-invasive techniques

With the microscope reactor it is possible to follow the development of micro-colonies individually in time. Figure 8 shows the development of a spherical colony in time. Cells expanded initially to spherical colonies. After some time (24 days) colonies appeared to be ellipsoid perpendicularly orientated to the surface of the gel. After 32 days the colony touched the surface and protruded slightly followed by an eruption after 36 days. After eruption a cavity remained.

Although the depth of field is not always perfect, because of the relatively large thickness of the gel plate (1 mm), it is sufficient to follow the expansion in time. The top of the photographs shows the liquid phase. It is difficult to observe a sharp liquid/gel interface (Hüsken et al. 1996).

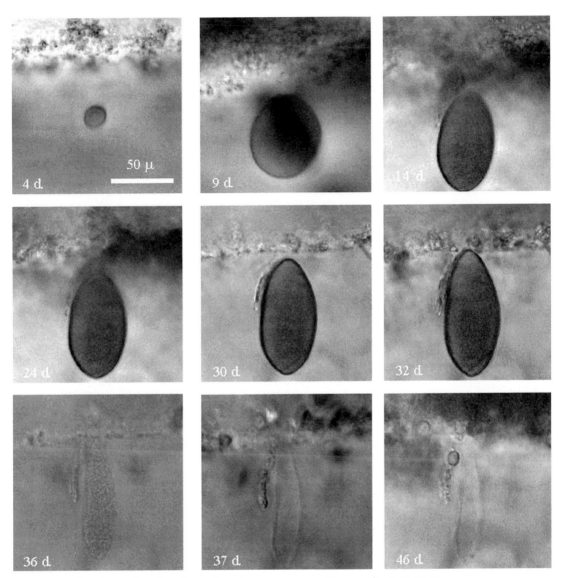

Fig. 8. Growth and eruption of a micro-colony in κ-carrageenan. Pictures after *4, 9, 14, 24, 30, 32, 36, 37* and *46* days of cultivation

Troubleshooting

For the conversion of the apparent colony radii to a relative biomass concentration, the following remarks apply. As the sections were not infinitely thin, the principle of Delesse was not directly applicable. Correction for the Holmes effect did lead to a 17 % lower volumetric fraction at a relative radius higher than 0.85. At a radius lower than 0.85, correction did lead to 44 % reduction. In the latter part of the bead however, only 10 % of the biomass was situated. During preparation of the beads shrinkage occurred. The radius decreased 16-19 %. We have assumed that there was no difference in shrinkage between colonies and beads. No protrusions were formed as the result of the fact that gel shrinkage was higher than colony shrinkage. There were no non-filled holes observed from which colonies had disappeared due to the fact that colony shrinkage was higher than gel shrinkage.

In the different methods different fixation techniques were used. There is no direct advice as to which one is the best. In general a fixative like paraformaldehyde penetrates the gel matrix faster than glutaraldehyde. Glutaraldehyde on the other hand is a stronger fixative. In the methods used, both fixatives gave good results.

When the protocol for the FA labelling was developed, autofluorescence of the aging colonies was a serious problem. For the suppression of this autofluorescence, blocking with NH_4Cl, $NaBH_4$ and BSA was introduced into the labelling protocol.

For the FA technique it is essential to work with infinitely thin slices because diffusion of antibodies in κ-carrageenan gel 3% w/v is negligible (Chevalier et al. 1987).

References

Al-Rubeai, M., Spier, R. (1989) Quantitative cytochemical analysis of immobilized hybridoma cells. Appl. Microbiol. Biotechnol. 31: 430-433

Karel, S.F., Robertson, C.R. (1989) Autoradiographic determination of mass-transfer limitations in immobilized cell reactors. Biotechnol. Bioeng. 34: 320-336

Boross, L., Papp, P. and Szajáni, B. Determination of the growth of gel-entrapped microbial cells at various depths of the alginate. In: Physiology of Immobilized Cells (De Bont, J.A.M., Visser, J., Mattiasson, B. and Tramper, J., eds.) Proceedings of an International Symposium held at Wageningen, The Netherlands, 10-13 December 1989. Elsevier Science Publishers B.V., Amsterdam, 1990, 201-204

Chevalier P, Cosentino GP, de la Noüe J, Rakhit S (1987) Comparative study on the diffusion of an IgG from various hydrogel beads. Biotechnol Techniques 1: 201-206

Chibata, I., Tosa, T. and Fujimura, F. (1983) Immobilized living microbial cells. In: Tsao, G.T. (ed.) Annual Reports on Fermentation Processes, vol. 6. Academic Press, London, p. 1-22

Chrzanowski, T.H., Crotty, R.D., Hubbard, J.G., Welch, R.P. (1984). Applicability of the fluorescein diacetate method of detecting active bacteria in freshwater. Microb. Ecol. 10: 179-185

Gosmann, B. and Rehm, H.J. (1986) Oxygen uptake of microorganisms entrapped in Ca-alginate. Appl. Microbiol. Biotechnol. 23, 163-167

Hennig A (1969) Fehler der Volumermittlung aus der Flächenrelation in dicken Schnitten (Holmes Effekt). Mikroskopie 25: 25-44

Heslop-Harrison J., Heslop-Harrison Y. (1970) Evaluation of pollen viability by enzymatically induced fluorescence; intracellular hydrolysis of fda. Stain Technol. 45: 115-120

Hunik, J.H., Van den Hoogen, M.P., De Boer, W., Smit, M., Tramper, J. (1993). Quantitative determination of the spatial distribution of *Nitrosomonas europaea* and *Nitrobacter agilis* cells immobilized in κ-carrageenan gel beads by a specific fluorescent-antibody labelling technique. Appl. Environ. Microbiol. 9: 1951-1954

Hunik, J.H., Bos, C.G., Van den Hoogen, M.P., De Gooijer, C.D., Tramper, J. (1994). Co-immobilized *Nitrosomonas europaea* and *Nitrobacter agilis* cells: validation of a dynamic model for simultaneous substrate conversion and growth in κ-carrageenan gel beads. Biotechnol. Bioeng. 43: 1153-1163

Hüsken L.E., Tramper J., Wijffels R.H. (1996) Growth and eruption of gel-entrapped microcolonies. In: R.H. Wijffels, R.M. Buitelaar, C. Bucke, J. Tramper (eds.) Immobilized cells: basics and applications. Elsevier Science BV, pp. 336-340

Karel, S.F., Robertson, C.R. (1989) Autoradiographic determination of mass-transfer limitations in immobilized cell reactors. Biotechnol. Bioeng. 34: 320-336

Khang, Y.H., Shankar, H. and Senatore, F. (1988) Modelling the effect of oxygen mass transfer on ß-lactam antibiotic production by immobilized *Cephalosporium acremonium*. Biotechnol. Lett. 10, 861-866

Kuhn, R.H., Peretti, S.W., Ollis D.F. (1991) Micro fluorimetric analysis of spatial and temporal patterns of immobilized cell growth. Biotechnol. Bioeng. 38: 340-352

Laanbroek H.J., Gerards S. (1991) Effects of organic manure on nitrification in arable soils. Biol Fertil Soils 12: 147-153

Leenen E.J.T.M., Boogert A.A., Van Lammeren A.A.M., Tramper J., Wijffels R.H. (1997) Dynamics of artificially immobilized *Nitrosomonas europaea*: effect of biomass decay. Biotechnol. Bioeng. 55: 630-641

Lefebvre, J. and Vincent, J.C. Dynamic simulations of cell-bearing membranes: modelling and optimization of bioreactors. European Symposium on Computer Aided Process Engineering 2, 5-7 October 1992, Toulouse, France, Supplement to Computers and Chemical Engineering 17, Pergamon Press, 1992, S221-S226

Monbouquette, H.G., Sayles, G.D. and Ollis, D.F. (1990) Immobilized cell biocatalyst activation and pseudo-steady-state behavior: model and experiment. Biotechnol. Bioeng. 35, 609-629

Underwood EE (1972) The stereology of projected images. J Microscopy 95: 25-44

Van Neerven A.R.W., Wijffels R.H., Zehnder A.J.B. (1990). Scanning electron microscopy of immobilized bacteria in gel beads: a comparative study of fixation methods. Journal of Microbiological Methods 11: 157-168

Verhagen F.J.M., Laanbroek H.J. (1991) Competition for ammonium between nitrifying and heterotrophic bacteria in dual energy-limited chemostats. Appl Environ Microbiol 57: 3255-3263

Wada, M., Kato, J. and Chibata, I. (1980) Continuous production of ethanol using immobilized growing yeast cells. Eur. J. Appl. Microbiol. Biotechn. 10, 275-287

Walsh, P.K., Brady, J.M. and Malone, D.M. Determination of the radial distribution of *Saccharomyces cerevisiae* immobilised in calcium alginate gel beads. Biotechnol. Techn. 1993, 7 (6), 435-440

Weibel ER (1979) Stereological methods 1, Academic Press, London, 415 p

Weibel ER (1980) Stereological methods 2, Academic Press, London, 340 p

Wijffels, R.H. and Tramper, J. (1989) Performance of growing *Nitrosomonas europaea* cells immobilized in k-carrageenan. Appl. Microbiol. Biotechnol. 32, 108-112

Wijffels, R.H., De Gooijer, C.D., Kortekaas, S. and Tramper, J. (1991) Growth and substrate consumption of *Nitrobacter agilis* cells immobilized in carrageenan: part 2. model evaluation. Biotechnol. Bioeng. 38: 232-240

Wijffels R.H., De Gooijer C.D., Schepers A.W., Beuling E.E., Mallée L.R., Tramper J. (1995) Growth of immobilized *Nitrosomonas europaea*: implementation of diffusion limitation over microcolonies. Enzyme and Microbial Technology 17: 462-471

Worden, R.M., Berry, L.G. (1992) The one-dimensional biocatalyst, a research tool for in situ analysis of immobilized-cell biocatalysts. Appl. Biochem. Biotechnol. 34/35: 487-498

Abbreviations

c	correction factor for the Holmes effect [-]
$E(D)$	expectation of observed diameter [m]
L	section thickness [m]
r_c	colony radius [m]
r_o	observed colony radius [m]
R	bead radius [m]
R_c	distance centre colony to centre of the bead [m]
R_t	radius testline [m]

NMR and Immobilized Cells

JEAN-NOËL BARBOTIN, JEAN-CHARLES PORTAIS, PAULA M. ALVES, and HELENA SANTOS

Introduction

Nuclear Magnetic Resonance (NMR) spectroscopy is based on the response of certain nuclei which possess an intrinsic magnetic moment (^1H, ^{13}C, ^{15}N, ^{19}F, ^{23}Na, ^{31}P, etc.) to an applied magnetic field. Such a technique can detect separate signals from various compounds within a sample and provide information on molecular identity and concentration. NMR is also widely recognized as a non-destructive, non-invasive technique for studying intracellular processes allowing *in vivo* estimation of intermediate metabolite concentrations in living systems. However, this is an insensitive non invasive method compared to most types of spectroscopy and requires very high cell densities for obtaining a good signal-to-noise ratio within a reasonable time.

The combination of immobilized cells and NMR spectroscopic methods is very convenient (Fernandez and Clark, 1987; Vogel et al, 1987) since high densities are obtained and maintained for extended periods of time under physiological conditions. Furthermore, NMR is also an effective method

✉ Jean-Noël Barbotin, UPRES-A CNRS 6022, Université de Picardie Jules Verne, Laboratoire de Génie Cellulaire, 33 rue Saint-Leu, Amiens, 80039, France (*phone* +33 3 22 82 75 95; *fax* +33 3 22 82 75 95; *e-mail* Jean-Noel.Barbotin@sc.u-picardie.fr)
Jean-Charles Portais, UPRES-A CNRS 6022, Université de Picardie Jules Verne, Laboratoire de Génie Cellulaire, 33 rue Saint-Leu, Amiens, 80039, France (*phone* +33 3 22 82 75 95; *fax* +33 3 22 82 75 95; *e-mail* Jean-Charles.Portais@sc.u-picardie.fr)
Paula M. Alves, Universidade Nova de Lisboa, Instituto de Tecnologia Química e Biológica, (ITQB), Rua da Quinta Grande 6, Apartado 127, Oeiras, 2780-156, Portugal (*phone* + 351-21-4469800; *fax* + 351-21-442-8766; *e-mail* marques@itqb.unl.pt)
Helena Santos, Universidade Nova de Lisboa, Instituto de Tecnologia Química e Biológica, (ITQB), Rua da Quinta Grande 6, Apartado 127, Oeiras, 2780-156, Portugal (*phone* + 351-21-4469800; *fax* + 351-21-442-8766; *e-mail* santos@itqb.unl.pt)

for determining diffusion coefficients of solutes or heavy metal uptake within the polysaccharidic gels used for cell immobilization.

State of the Art

Animal Cells

The first attempt, using ^{31}P NMR (the 31-phosphorus nucleus is 100% abundant), was reported by Ugurbil et al (1981) and concerned anchorage-dependant mouse embryo fibroblasts immobilized in Cytodex microcarrier beads. Gonzalez-Mendez et al (1982) have employed a hollow fiber system mounted inside an NMR tube to record ^{31}P spectra of immobilized hamster cells. Knop et al (1984) and Foxall et al (1984) monitored ATP levels in Chinese hamster lung fibroblasts immobilized in agarose gelthreads. Hvorat et al (1985) further generalized the use of hollow fiber membrane reactors (HFBR) for NMR cell culture by constructing a hollow fiber cartridge adaptable to a variety of standard NMR probes. Some NMR flow imaging techniques have been developed to measure fluid flow in a cell free HFBR with a view to optimizing the operating conditions and for improving the design of such devices (Pangrle et al, 1989, Hammer et al, 1990, Yao et al, 1995). ^{13}C NMR has been used to observe the intracellular environment of mammalian cells (Fernandez et al, 1990, Mancuso et al, 1994) and quite recently Mancuso et al (1998) have demonstrated that changes in extracellular glutamine concentration can affect primary and secondary metabolism of murine hybridomas in a HFBR. ^{31}P and ^{13}C NMR were also used to study the energy metabolism in perfused human erythrocytes immobilized in agarose and alginate threads (Lundberg et al, 1992). Glioma cells and primary astrocytes immobilized in microcarriers have been used to study the effect of anoxia, ischaemia and glucose starvation on the cellular energetic *status* (Pianet et al, 1991). A similar immobilization strategy was followed to study cellular calcium homeostasis in neuroblastoma cells using ^{19}F-NMR spectroscopy (Benters et al, 1997). Regulation of intracellular pH in neuronal and glial tumour cells was studied by ^{31}P in cells entrapped in basement membrane gel (BMG) threads (Flögel et al. 1994). Primary brain neurons and astrocytes from rat have been immobilized in porous microarriers or BMG threads and studied by ^{31}P-NMR (Alves et al, 1996, Alves et al, 1997). Changes in the pools of organic solutes, cell volume and other metabolic parameters associated with osmotic stress were studied by multinuclear *in vivo* NMR (^{31}P and diffusion-weighted ^{1}H) in a glial cell line immobilized in gel threads (Flögel et al, 1995). Hepatocytes immobilized within agarose beads have been studied by ^{31}P NMR (Farghali et al, 1992) to test toxicity. On the other hand, oxygenation is a major determinant of the physiological state of cultured cells. ^{19}F-NMR has been used to determine the oxygen concentration available to tumor cells immobilized in alginate gel beads by measuring the relaxation rate ($1/T_1$) of perfluorocarbons incorporated into the gel matrix (McGovern et al, 1993). A membrane oxygenator has been described for the perfusion of cultured cells (Gamcsik et al, 1996) and it can be positioned at the opening of the magnet bore which allows oxygenation.

Recently, Doliba et al (1998) developed an NMR method to study on-line mitochondrial function. Mitochondria were maintained in a stable physiological state in agarose beads that were continuously superfused with oxygenated buffer and ATP production was observed to be very sensitive to hypoxia and ischaemia. Ojcius et al (1998) reported the use of *in vivo* ^{31}P- and ^{13}C-NMR to study metabolic changes associated with infection by *Chlamydia* of HeLa cells immobilized in microspheres.

In vivo studies of free and immobilized cells of *Catharanthus roseus* have been investigated by Vogel and Brodelius (1984). The cells were cultivated in submerged cultures in a modified growth medium (low concentration of manganese, which is paramagnetic and therefore can interfere in the NMR experiments) and they were entrapped in alginate or agarose according to standard procedures. This data demonstrated that there were no differences in many of the relevant intracellular regulatory parameters (ATP and P_i levels, NAD(P), etc.) between immobilized and freely suspended cells.

Plant Cells

A similar behaviour between free and immobilized cells has been reported by Brindle and Krikler (1985) with embedded yeasts in agarose gel threads that were continuously perfused with fresh nutrient medium throughout the NMR experiments. However, Galazzo et al (1987) and Galazzo and Bailey (1989), using always ^{31}P NMR, have noticed differences in intracellular phosphate chemical shift between free and immobilized yeasts (Ca-alginate beads). These authors stated that the intracellular pH in immobilized cells was lower than in cell suspensions. By using ^{13}C NMR and ^{13}C-labelled glucose they have shown an increase in polysaccharide production by immobilized yeasts. Some metabolic studies have also been performed by using ^{31}P NMR with a unicellular algae (*Dunaliella salina*) entrapped in agarose beads (Bental et al, 1990) showing a stable and active state of the immobilized cells. Briasco et al (1990 a and b) have demonstrated the presence of significant mass transfer limitations in a perfused *E. coli* system (HFBR) during NMR spectroscopy. The NMR spectra obtained from this reactor showed near-normal intracellular pH, metabolite concen-

Microbial Cells

trations and NTP/NDP ratios. The basis for the inhibition of denitrification by nitrite accumulated endogenously by *Pseudomonas fluorescens* immobilized in κ-carrageenan have been investigated by ^{31}P-NMR (Sijbesma et al, 1996). The same immobilization system was used to investigate pH homeostasis in *Desulfovibrio gigas* by ^{31}P-NMR (Santos et al, 1994). A comparison between the anaerobic metabolism of glucose by suspended and κ-carrageenan entrapped *Saccharomyces cerevisiae* was reported by Taipa et al (1993). Lohmeier-Vogel et al (1995, 1996) used ^{31}P- and ^{13}C-NMR to study the capacity of agarose immobilized cultures of *Candida tropicalis*, *Pichia stipitis* and *Saccharomyces cerevisiae* to metabolize glucose or xylose.

Diffusion in Gels

The conformation of the basic components of gels used for cell immobilization can be investigated by using NMR methods. For example, Steginsky et al (1992) have studied the interaction of alginate constituents with calcium ion. NMR studies have also been developed to study the influence of potassium ions on sol-gel transitions with dextrans and carrageenans (Watanabe et al, 1996). Otherwise, diffusion coefficients are a fundamental measure to probe the diffusive resistance to gel beads with or without entrapped cells. Balcom et al (1993) have described a simple method to measure diffusion coefficients of paramagnetic species in aqueous gel media using one-dimensional NMR imaging. Diffusion coefficients of both organic (4-amino-TEMPO) and inorganic (copper sulfate) paramagnetic tracers in 10% polyacrylamide, 1% agarose have been reported by these authors. Diffusion and relaxation of solutes like glucose, glycine, lactate, sodium ions have been studied in agarose and alginate gels by using ^{1}H and ^{23}Na pulsed field gradient spin echo NMR techniques (Lundberg et Kuchel, 1997). NMR imaging studies have also been used to study the gelation of sodium alginate with calcium ions (Potter et al, 1994). This method provided the possibility to study heavy metal absorption in alginate and immobilized cells biosorbents (Nestle and Kimmich, 1996a) and to directly monitor the time evolution of the spatial distribution of the ions in the materials. Using Cu, an intrusion of only about 2 mm in 10 h was demonstrated (Nestle and Kimmich, 1996b). Recently, Degrassi et al (1998) reported a fully automated magnetic resonance imaging procedure to measure all the MRI parameters of water in both sodium alginate solutions and calcium alginate gels at concentrations from 1 to 4% (w/w). The spin/spin T_2 relaxation time and the proton exchange rates appeared to give useful information about molecular motion and gelling mechanisms. The mobility of water in intact biofilms was also measured with pulsed field gradient nuclear magnetic resonance and used to characterize their diffusive properties (Beuling et al, 1998). However, if in such a case some supplementary information on cell fraction and spatial organization could be obtained, quantitative analysis was not yet possible.

Materials

A wide range of materials suitable for cell immobilization and NMR spectroscopy are commercially available. Agarose, carrageenan and alginate are often used for immobilization of microbial or plant cells. Agarose (low-melting-temperature), basement membrane gel (mainly laminin and collagen), microcarriers of cross-linked dextran with several surface treatments (Cytodex), gelatin macroporous beads (CultiSpher), macroporous cellulose cubic supports with different coating materials (Cellsnow), polystyrene (Biosylon), and several hollow-fiber systems are preferred matrices for immobilization of animal cells.

To select an immobilization system for *in vivo* NMR spectroscopy the following aspects should be taken into consideration:

- the immobilization material has to be chemically inert and should not interfere with cell metabolism;

- the material should not give rise to strong NMR signals that may obscure resonances due to cell metabolites;

- the immobilized system should be easy to handle in sterile conditions;

- the proportion of the detection volume occupied by the immobilization matrix should be small;

- movements of the immobilized packed material may perturb NMR measurements, especially in studies involving imaging techniques;

- the mechanical strength of the matrix should allow extended periods of time under perfusion without significant deterioration.

Procedure

Ugurbil and co-workers (1981) were the first to report a perfusion system for *in vivo* NMR spectroscopy that involved cell growth on the surface of microspheres contained in a perfusion chamber. This procedure is applicable to anchorage-dependant cells, but the final cell density is determined by the bead size. A major disadvantage of this system derives from the large proportion of the detection volume occupied by the solid support. Therefore, low cell densities are obtained and long acquisition times are required. In addition, cells are subjected to shear stress due to perfusion which may perturb their metabolism or cause wash-out. As a consequence, only low perfusion rates are permitted. Also, extensive growth often occurs in the inter-bead region that may cause mass transfer limitations in these areas and lead to overpressure in the NMR tube (or chamber). These problems can be overcome by using three-dimensional matrices for cell immobilization. The main advantages of these systems are their capability of supporting a higher cell density and allowing cell growth in an environment protected from shear damage; higher perfusion rates can be used that avoid nutrient and oxygen limitations as well as accumulation of waste products.

Gel matrices (threads or beads) and porous microcarriers are attractive alternatives for three-dimensional structures, and are the only approach suitable to immobilize cells capable of growth in suspension. They have the advantages of allowing faster rates of perfusion and causing no appreciable build-up of back-pressure. For cells with high growth rates, porous microcarriers are recommended because they are reasonably priced and easy to handle and transfer. Main disadvantages are associated with the inefficiency of the inoculation process that should be able to force cells entering the porous matrix, and with the impossibility of direct cell visualization. In contrast, losses during inoculation are drastically reduced by embedding the cells in a permeable matrix (for instance, gel threads) because cells are actually entrapped during the gel polymerization, and consequently higher cell densities are obtained.

Immobilization systems are essential in experiments where NMR is used to monitor metabolism along cell growth; in this case, porous microcarriers are preferred due to their high resistance to mechanical stress as compared, for instance, with gel matrices. Furthermore, the perfusing flow is used to assist trapping the cells inside the pores during the inoculation process (Thelwall and Brindle, 1999).

1 Immobilization of animal cells for *in vivo* NMR spectroscopy

1. 1 Immobilization in gel matrices

For cell immobilization, gel can be shaped into either thin threads or small diameter beads. Threads are generally preferred since small beads are more difficult to obtain. However, for identical thickness, their resistance to mechanical damage is lower and careful manipulation is required to avoid disruption of thin threads.

1.1.1 Basement membrane gel (BMG)

This matrix is particularly suited for immobilization of both primary cells and cell lines. High cost is the only disadvantage worth mentioning.

Materials

Matrigel solution is obtained from Serva (Heidelberg, Germany); Dulbecco's modified Eagle's medium (DMEM), foetal bovine serum (FBS), phosphate-buffered saline (PBS) and trypsin-EDTA are from GIBCO (Glasgow, U.K.). T-flasks (175 cm^2) for animal cell cultures and bacteriological Petri dishes (145 cm^2) are from Nunc (Roskilde, Denmark). Sterile syringes and Teflon tubing (300 µm inner diameter) are from regular commercial sources. All the material in contact with the gel has to be kept at 4°C.

Cell culture

Cells are cultured in T-flasks until they reach confluence or until the desired growth stage is reached, using typical procedures for animal cell culture (Doyle et al, 1996). For a 10 mm NMR tube, approximately 10^7 cells are needed. Cells are washed once with PBS and harvested from the T-flask surface by trypsinization. After centrifugation (300 x *g*, 10 minutes) the cell pellet is resuspended in 1 ml of DMEM. Note that the inoculum size can be increased or decreased depending on cell type and on the purpose of the experiment.

Immobilization

Cells are embedded in the gel by gently mixing the cell suspension with 1 ml of liquid BMG at 4°C, using a micro-pipette with a frozen tip. The whole procedure is carried out on ice. The mixture is collected into sterile syringes (1 ml volume) and placed at 37°C for 5 minutes. A Teflon tube (about 10 cm long) is then fitted to the syringe and the gel is slowly extruded into Petri dishes containing DMEM at 37°C. This operation has to be performed carefully in order to avoid the disruption of the newly formed threads. Petri dishes suitable for bacterial growth should be used because their surfaces prevent the adhesion of animal cells. Threads are placed in the incubator, at 37°C in a humidified atmosphere of 5% CO_2 in air and kept for 3-5 days to allow cell growth. Growth inside the threads can be viewed using an inverted microscope. This procedure is adapted from Daly et al, 1987 and Alves et al, 1996.

1.1.2. Agarose

Materials Low-melting agarose from Sigma, Dulbecco's modified Eagle's medium (DMEM), foetal bovine serum (FBS), phosphate-buffered saline (PBS) and trypsin-EDTA are from GIBCO (Glasgow, U.K.). T-flasks (175 cm^2) for animal cell cultures are from Nunc (Roskilde, Denmark). Sterile syringes and Teflon tubing (500 µm inner diameter) are from regular commercial sources.

Cell culture Cells are prepared as described above with the exception that threads are used for NMR analysis immediately after the preparation. Cell density is usually in the order of 5 x 10^7 cell/ml.

Immobilization Agarose is dissolved at 70°C in Krebs-Henseleit solution (NaCl 6.99 g/l, KCl 0.36 g/l, KH_2PO_4 0.13 g/l, $MgSO_4 \cdot 7H_2O$ 0.295 g/l, $CaCl_2 \cdot H_2O$ 0.37 g/l, $NaHCO_3$ 2 g/l, pH 7.4) to a concentration of 22 mg/ml. The agarose solution is equilibrated to 37°C in a water bath for approx. 10 min. Equal volumes of cell resuspension and agarose are mixed with a pipette. The mixture is sucked into a syringe which is placed at 37°C. The mixture is then forced through the Teflon tubing at a rate of 0.5-1 ml/min. Approximately 0.5 m of Teflon tube is placed in an ice/water mixture to bring about gelatinization of the agarose. The threads are collected in perfusion medium contained in a 10 mm NMR tube held at a temperature of approx. 15°C. Threads are compressed gently to the final volume in the NMR tube with a plunger. This protocol is adapted from Lundberg et al, 1994 and Foxal and Cohen, 1983.

1.2. Immobilization in microcarriers

The immobilization procedure depends on the matrix composition of the microcarrier, as well as on its shape and porosity. Usually, manufacturers recommend specific inoculation procedures for each type of carrier. Here, we only consider microcarriers that are suitable for NMR analysis and we describe the protocol to immobilize an anchorage-dependent animal cell line (Baby Hamster Kidney, BHK) in two non-porous (Cytodex 3 from Pharmacia; Biosylon from Nunc) and two porous microcarriers (Culti-Spher S from Percell; Cellsnow EX from Kirin).

Dulbecco's modified Eagle's medium (DMEM), foetal bovine serum (FBS), phosphate-buffered saline (PBS) and trypsin-EDTA are obtained from GIBCO (Glasgow, U.K.). T-flasks (175 cm^2) for animal cell cultures and bacteriological Petri dishes (145 cm^2) are obtained from Nunc (Roskilde, Denmark). Spinner flasks are obtained from Corning (USA). Cytodex 3 is a microsphere of a cross-linked dextran matrix covered by a thin layer of denatured collagen from Pharmacia. Biosylon is a microsphere of polystyrene from Nunc. Culti-Spher is a spherical macroporous gelatin microcarrier from Percell and Cellsnow is a cubic macroporous cellulose carrier from Kirin.

Materials

Cells are prepared as described above for immobilization in BMG threads with the exception that 5 ml of DMEM, instead of 1 ml, are used for the final resuspension of the cell pellet.

Cell culture

Microcarriers are allowed to swell overnight in PBS without calcium and magnesium. After sterilization by autoclaving, PBS is removed and 2.5 ml of microcarriers are placed in sterile vials and covered with DMEM containing 10% FBS (the serum is used to promote cell adherence). The culture medium is removed prior to inoculation.

Immobilization

The cell suspension is spread on the top of the carriers using a pipette. The vials are placed in the incubator (37°C, humidified atmosphere of 5% CO$_2$ in air) for 3 hours with occasional gentle manual agitation. The inoculated supports are then transferred to four bacteriological Petri dishes containing 25 ml of culture medium (DMEM with 10% FBS), and allowed to grow for 5-7 days until they reach confluence or the desired growth stage.

Non-porous microcarriers

The cell suspension is forced into the carriers by sprinkling under vigorous manual agitation. The vials are placed in the incubator (37°C, humidified atmosphere of 5% CO$_2$ in air) and agitated gently every 30 minutes for a period of 3 hours. After this time the microcarriers are transferred to a spinner flask containing 50 ml culture medium and agitated at 30 rpm. Twenty-four hours after immobilization, 50 ml of culture medium are added to the spinner vessel and the agitation rate is increased to 60 rpm to avoid mass transfer limitations. Cells are allowed to grow for 4-5 days and, whenever necessary, fresh culture medium is supplied.

Porous microcarriers

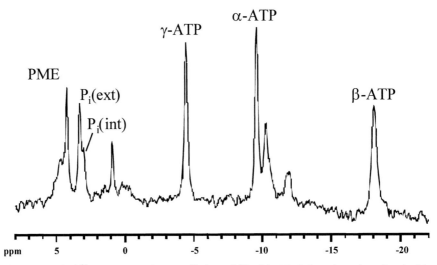

Fig. 1. A typical ^{31}P spectrum of BHK cells immobilized in BMG threads and perfused with DMEM at 1.5 ml/min (10 min. acquisition). The spectrum was acquired in a Bruker DRX500 spectrometer. Abbreviations: PME, phosphomonoesters; Pi(ext), external inorganic phosphate; Pi(int), internal inorganic phosphate

1.3. Perfusion

For NMR experiments, the immobilized cells are collected and transferred to a sterile NMR chamber fitted with a perfusion insert. The whole perfusion system has been previously sterilized by autoclaving. Typical perfusion rates are in the range of 1 to 10 ml/min. Cells are perfused with culture medium or buffer (usually DMEM or PBS at 37°C) continuously gassed with 95%O_2/5%CO_2. A typical ^{31}P-NMR spectrum of BHK cells immobilized in BMG threads is shown in Figure 1 (10 min. acquisition).

2. Immobilization of microbial cells for *in vivo* NMR spectroscopy

Several methods for immobilization of microbial cells with preserved viability have been described. (Nilson et al, 1983; Sijbesma et al, 1996, Taipa et al, 1993, Santos et al, 1994, Brindle and Krikler, 1985, Galazzo et al, 1987). Here we describe protocols that can be followed to immobilize different microbial cells for *in vivo* NMR spectroscopy.

2.1. Immobilization (Threads or Beads)

κ-carrageenan, low-gelling temperature agarose, and alginate are obtained from Sigma (Deisenhofen, Germany).

Materials

Cells are harvested by centrifugation (e.g. 5000 x g for 9 minutes), washed with an appropriate buffer solution, and resuspended in the same solution.

For cell immobilization 0.072 g of κ-carrageenan is dissolved in 3.3 ml of distilled water at 90°C. This solution is allowed to cool down to room temperature and is then mixed with ca. 1 g (wet weight) of freshly prepared cell paste suspended in 1.1 ml of physiological saline solution. The mixture is then extruded, via an 18-gauge needle connected to a flexible plastic tube, through a microcapillary pipette (0.16 mm internal diameter) into an ice-cold 0.3 mM KCl solution. Thin gel-threads (0.3 mm average diameter) are usually obtained in this manner. After allowing to cure for 20 minutes, threads are washed with buffer, suspended in the same buffer and immediately used for NMR. This procedure is adapted from Sijbesma et al, 1996.

κ-carrageenan

For cell immobilization 4 g of cell pellet are warmed to 30-35°C and mixed with 4 ml of 1.8% low-gelling temperature agarose which is dissolved in growth medium or physiological saline solution. The cell/agarose mixture is extruded in 1 ml aliquots through approx. 0.4 m of 500 μm internal diameter flexible plastic tubing which is cooled in an ice/water bath to bring about gelation of the agarose. The resulting gel threads are collected in perfusion medium contained in a NMR tube and immediately used for NMR experiments. This method is adapted from Brindle and Klikler, 1985 and Foxal and Cohen, 1983.

Agarose

To immobilize cells in alginate beads, the pellet is resuspended in medium and mixed with the same volume of 4% Na-alginate. The homogeneous mixture is filled into a cylindric reservoir and pressurized nitrogen gas is used to force the solution through a Teflon tube (0.8 mm inner diameter). A coaxial tube with a larger diameter (1.8 mm) is used to blow off the alginate droplets into 2% $CaCl_2$ solution at a controlled size by a compressed concentric air stream. Beads are allowed to cure for 0.5 h, then washed in resuspending medium and used for the NMR experiments. Beads of diameters varying between 0.1 and 2 mm can be made. The bead size can be adjusted by varying the nitrogen flow rate. These procedures are described in Gallazo and Bailey, 1990.

Alginate

3. Immobilization of plant cells for *in vivo* NMR spectroscopy

Protocols for immobilization of microbes can be easily adapted for immobilization of plant cells (Vogel et al., 1987)

References

Alves PM, Carrondo MJT, Santos, H (1997) Immobilization of primary brain cells in porous microcarriers for on line NMR spectroscopy in Carrondo MJT, Griffiths B, Moreira JL (eds.) Animal Cell Technology, from Vaccines to Genetic Medicine. Kluwer Academic Publishers, London, pp 91-98

Alves PM, Flögel U, Brand A, Leibfritz D, Carrondo MJT, Santos H, Sonnewald U (1996) Immobilization of primary astrocytes and neurons for NMR monitoring of biological processes in vivo. Dev Neurosci 18:478-483

Balcom BJ, Fischer AE, Carpenter TA, Hall LD (1993) Diffusion in aqueous gels. Mutual diffusion coefficients measured by one-dimensional Nuclear Magnetic Resonance Imaging. J Am Chem Soc 115:3300-3305

Bental M, Pick U, Avron M, Degani H (1990) Metabolic studies with NMR spectroscopy of the alga *Dunaliella salina* trapped within agarose beads. Eur J Biochem 188:111-116

Benters J, Flogel U, Schafer T, Leibritz D, Hechtenberg S, Beyersmann D (1997) Study of the interactions of cadmium and zinc ions with cellular calcium homeostasis using ^{19}F-NMR spectroscopy. Biochem J 322:793-799

Beuling EE, van Dusschoten D, Lens P, van den Heuvel JC, Van Has H, Ottengraf SSP (1998) Characterization of the diffusive properties of biofilms using pulsed field gradient-nuclear magnetic resonance. Biotechnol Bioeng 60:283-291

Briasco CA, Karel SF, Robertson CR (1990) Diffusional limitations of immobilized *Escherichia coli* in hollow-fiber reactors: influence on ^{31}P NMR spectroscopy. Biotechnol Bioeng 36:887-901

Briasco CA, Ross DA, Robertson CR (1990) A hollow-fiber reactor design for NMR studies of microbial cells. Biotechnol Bioeng 36:879-886

Brindle K, Krikler S (1985) ^{31}P NMR saturation transfer measurements of phosphate consumption in *S. cerevisiae* . Biochim Biophys Acta 847:285-292

Daly PF, Lyon RC, Straka EJ, Cohen JS (1987) ^{31}P-NMR Spectroscopy of human cancer cells proliferating in a basement membrane gel. FASEB J. 2:2596-2604

Degrassi A, Toffanin R, Paoletti S, Hall LD (1998) A better understanding of the properties of alginate solutions and gels by quantitative magnetic resonance imaging (MRI). Carbohyd Res 306:19-26

Doliba NM, Wehrli SL, Babsky AM, Doliba NM, Osbakken MD (1998) Encapsulation and perfusion of mitochondria in agarose beads for functional studies with ^{31}P-NMR. Magnet Reson Med 39:679-684

Doyle A, Griffiths J, Newell D (eds) (1996) Cell & Tissue Culture: Laboratory Procedures; Basic techniques for primary cell cultures and establishment of continuous cultures. John Wiley & Sons, Chichester

Farghali H, Rossaro L, Gavaler J, van Thiel D, Dowd S, Williams D, Ho C (1992) ^{31}P-NMR spectroscopy of perfused rat hepatocytes immobilized in agarose threads: application to chemical-induced hepatotoxicity. Biochim Biophys Acta 1139:105-114

Fernandez EJ, Clark DS (1987) N.m.r. spectroscopy: a non-invasive tool for studying intracellular processes. Enzyme Microb Technol 9:259-271

Fernandez EJ, Mancuso A, Murphy MK, Blanch HW ans Clark DS (1990) Nuclear magnetic resonance methods for observing the intracellular environment of mammalian cells. Ann N Y Acad Sci 589:458-475

Flögel U, Niendorf T, Serkova N, Brand A, Henke J, Leibfritz D (1995) Changes in organic solutes, volume, energy state and metabolism associated with osmotic stress in a glial cell line: a multinuclear NMR study. Neurochem Res 20:793-802

Flögel U, Wilker W, Leibfritz D (1994) Regulation of intracellular pH in neuronal and glial tumour cells, studied by multinuclear NMR Spectroscopy. NMR Biomed 7:157-166

Foxall DL, Cohen JS (1983) NMR studies of perfused cells. J Magn Reson. 52:346-349

Foxall DL, Cohen JS, Mitchel JB (1984) Continuous perfusion of mammalian cells embedded in agarose gel threads. Exp Cell Res 154:521-529

Galazzo JL, Bailey JE (1989) In vivo nuclear magnetic resonance analysis of immobilization effects on glucose metabolism of yeast Saccharomyces cerevisiae. Biotechnol Bioeng 33:1283-1289

Galazzo JL, Bailey JE (1990) Growing Saccharomyces cerevisiae in calcium-alginate beads induces cell alterations which accelerate glucose conversion to ethanol. Biotechnol Bioeng 36:417-426

Galazzo JL, Shanks JV, Bailey JE (1987) Comparison of suspended and immobilized yeast metabolism using ^{31}P nuclear magnetic resonance spectroscopy. Biotechnol Techn 1:1-6

Gamcsik MP, Forder JR, Millis KK, McGovern KK (1996) A versatile oxygenator and perfusion system for magnetic resonance studies. Biotechnol Bioeng 49:348-354

Gonzalez-Mendez R, Wemmer D, Hahn G, Wade-Jardetzky N, Jardetzky O (1982) Continuous flow NMR culture system for mammalian cells. Biochim Biophys Acta 720:274-280.

Hammer BE, Heath CA, Mirer SD, Belfort G (1990) Quantitative flow measurements in bioreactors by nuclear magnetic resonance imaging. Bio/Technol 8:327-330

Hrovat ML, Wade CG, Hawkes SP (1985) A space-efficient assembly for NMR experiments on anchorage-dependant cells. J Magn Reson 61:409-417

Kaplan O, Cohen JS (1994) Lymphocyte activation-^{31}P magnetic resonance studies of energy metabolism and phospholipid pathways. ImmunoMethods. 4:163-178

Knop RH, Chen CW, Mitchell JB, Russo A, McPherson S, Cohen JS (1984) Metabolic studies of mammalian cells by ^{31}P-NMR using a continuous perfusion technique. Biochim Biophys Acta 804:275-84

Lohmeier EM, Hahn-Hägerdal B, Vogel HJ (1995) Phosphorus-31 and Carbon-13 nuclear magnetic resonance study of glucose and xylose metabolism in agarose-immobilized Candida tropicalis. Appl Environ Microbiol 61:1420-1425

Lohmeier EM, McIntyre DD, Vogel HJ (1996) Phosphorus-31 and Carbon-13 nuclear magnetic resonance studies of glucose and xylose metabolism in cell suspensions and agarose-immobilized cultures of Pichia stipitis and Saccharomyces cerevisiae Appl Environ Microbiol 62:2832-2838

Lundberg P, Kuchel PW (1994) Immobilization methods for NMR studies of cellular metabolism - A practical guide. ImmunoMethods. 4:163-178

Lundberg P, Kuchel PW (1997) Diffusion of solutes in agarose and alginate gels: ^{1}H and ^{23}Na PFGSE and ^{23}Na TQF NMR studies. Magnet Reson Med 37:44-52

Lundberg P, Berners-Price SJ, Roy S, Kuchel PW (1992) NMR studies of erythrocytes immobilized in agarose and alginate gels. Magnet Reson Med 25:273-288

Mancuso A, Sharfstein ST, Fernandez EJ, Clark DS, Blanch HW (1998) Effect of extracellular glutamine concentration on primary and secondary metabolism examination of a murine hybridoma: an in vivo ^{13}C nuclear magnetic resonance study. Biotechnol Bioeng 57:172-186

Mancuso A, Sharfstein ST, Tucker SN, Clark DS, Blanch HW (1994) Examination of primary metabolic pathways in a murine hybridoma with carbon-13 nuclear magnetic resonance spectroscopy. Biotechnol Bioeng 44:563-585

McGovern KA, Schoeniger JS, Wehrle JP, Ng CE, Glickson JD (1993) Gel-entrapment of perfluorocarbons: a fluorine-19 NMR spectroscopic method for monitoring oxygen concentration in cell perfusion systems. Magnet Reson Med 29:196-204

Nestle N, Kimmich R (1996a) NMR microscopy of heavy metal absorption in calcium alginate beads. Appl Biochem Biotechnol 56:9-17

Nestle N, Kimmich R (1996b) NMR imaging of heavy metal absorption in alginate, immobilized cells, and kombu algal biosorbents. Biotechnol Bioeng 51:538-543

Nilson K, Birnbaum S, Flygare S, Linse L, Schröder U, Jeppsson U, Larsson P-O, Mosbach K, Brodelius P (1983) A general method for the immobilization of cells with preserved viability. Eur J Appl Microb Biotechnol. 17:319-326

Ojcius DM, Degani H, Mispelter J, Dautry-Varsat A (1998) Enhancement of ATP levels and glucose metabolism during an infection by *Chlamydia*, NMR studies of living cells. J Biol Chem 273:7052-7058

Pangrle BJ, Walsh EG, Moore S, DiBiasio (1989) Investigation of fluid flow patterns in a hollow fiber module using magnetic resonance velocity imaging. Biotechnol Techn 3:67-72

Pianet I, Merle M, Labouesse J, Canioni P (1991) Phosphorus-31 nuclear magnetic resonance of C6 glioma cells and rat astrocytes, evidence for a modification of the longitudinal relaxation time of ATP and Pi during glucose starvation. Eur J Biochem 195:87-95

Potter K, Balcom BJ, Carpenter TA, Hall LD (1994) The gelation of sodium alginate with calcium ions studied by magnetic resonance imaging (MRI). Carbohydr Res 257:117-126

Santos H, Fareleira P, LeGall J, Xavier AV (1994) In vivo nuclear magnetic resonance in study of physiology of sulfate-reducing bacteria. Method Enzymol 243:543-558

Sijbesma WFH, Almeida JS, Reis MAM, Santos H (1996) Uncoupling effect of nitrite during denitrification by *Pseudomonas fluorescens*: an in vivo ^{31}P-NMR study. Biotechnol Bioeng 52:176-182

Steginsky CA, Beale JM, Floss HG, Mayer RM (1992) Structural determination of alginic acid and the effects of calcium binding as determined by high-field n.m.r. Carbohydr Res 225:11-26

Taipa MA, Cabral JMS, Santos H (1993) Comparison of glucose fermentation by suspended and gel-entrapped yeast cells: an in vivo nuclear magnetic resonance study. Biotechnol Bioeng 41:647-653

Thelwall PE, Brindle KM (1999) Analysis of CHO-K1 cell growth in a fixed bed bioreactor using magnetic resonance and imaging. Cytotechnology 31:121-132

Ugurbil K, Guernsey DL, Brown JR, Glynn P, Tobkes N, Edelman IS (1981) ^{31}P NMR studies of intact anchorage-dependant mouse embryo fibroblasts. Proc Natl Acad Sci USA 78:4843-4847

Vogel HJ, Brodelius P (1984) An in vivo ^{31}P-NMR comparison of freely suspended and immobilized *Catharanthus roseus* plant cells. J Biotechnol 1:159-170

Vogel HJ, Brodelius P, Lilja H, Lohmeier-Vogel EM (1987) Nuclear magnetic resonance studies of immobilized cells. Method Enzymol 135:512-528

Watanabe T, Ohtsuka A, Murase N, Barth P, Gersonde K (1996) NMR studies on water and polymer diffusion in dextran gels. Influence of potassium ions on microstructure formation and gelation mechanism. Magnet Reson Med 35:697-705

Yao S, Costello M, Fane AG, Pope JM (1995) Non-invasive observation of flow profiles and polymerisation layers in hollow fibre membrane filtration modules using NMR micro-imaging. J Membrane Sci 99:207-216

Glossary

Chemical shift	the immediate chemical environment of a nucleus in a molecule induces slight shifts of its resonance (or Larmor) frequency (10^6 order of magnitude lower than the spectrometer frequency). The NMR spectra are then conveniently scaled in chemical shift (δ), defined as follows: $\delta = 10^6 \cdot (\nu_0 - \nu_{ref})/\nu_{ref}$ where ν_0 is the Larmor frequency of the nucleus, ν_{ref} the Larmor frequency of a reference. Note that the chemical shift is independant of the spectrometer frequency.
Magnetic Resonance Imaging	a NMR technique for obtaining images based on the spatial discrimination of a nucleus (generally the water protons) by using magnetic field gradients instead of a static magnetic field.
One-dimensional (1D) NMR	single NMR experiments in which the sample nuclei are excited only once before the signal acquisition. This signal (a free induced decay or FID) is converted through a Fourier-Transform into a spectrum.
Pulsed field gradient spin echo	a NMR method based on the spatial discrimination of nuclei in the sample volume. This method allows the monitoring of molecular diffusion and hence the measurement of the diffusion coefficients.
Relaxation	refers to the return of an excited nucleus back to its fundamental state. Two types of relaxation rates are distinguished: the longitudinal relaxation rate ($1/T_1$), which characterizes the evolution of the nuclear magnetic moment along the axis defined by the magnetic field, and the transversal relaxation ($1/T_2$), which characterizes the evolution of the nuclear magnetic moment in the plane perpendicular to the magnetic field axis.

Two-dimensional (2D) NMR serial NMR experiments in which the sample nuclei are excited several times at incremented intervals. The series of FIDs are usually converted into a map (contour plot). Such spectroscopy is used to evidence correlations between two (or more) different nuclei.

Immobilization at Large Scale by Dispersion

DENIS PONCELET, STEPHANE DESOBRY, ULRICH JAHNZ, and K. VORLOP

Introduction

Producing microcapsules at large scale implies first caring about the dispersion processes. The droplet solidification step (gelation, membrane formation ...) generally brings less problems (reader will refer to Chapter 3). In this chapter, four technologies of producing droplet at large scale will be proposed for cell encapsulation purposes:

- jet breaking methods

- LentiKat method

- rotating devices

- emulsification using static mixers

Considering that scale-up is performed in industries, information about large scale equipment remains limited. However, this chapter will try to provide as much complete technical information as possible on the subject.

✉ Denis Poncelet, ENITIAA, Rue de la Géraudiere, BP 82 225, Nantes, 44 322, France (*phone* +33-251-785425; *fax* +33-251-785467; *e-mail* poncelet@enitiaa-nantes.fr; *homepage* BRG.enitiaa-nantes.fr)
Stephane Desobry, ENSAIA, BP 172, Vandoeuvres-les-Nancy, 54505, France
Ulrich Jahnz, geniaLab, BioTechnologie, Bundesallee 50, Braunschweig, 38116, Germany
K. Vorlop, FAL Institut fuer Technologie, Bundesallee 50, Braunschweig, 38116, Gemany

Subprotocol 1
Jet Breaking Methods

Principle

While liquid is extruded at large flow (› 30 ml/h) from a nozzle, the formed liquid jet can be broken into droplets of narrow size distribution by two methods: the nozzle resonance method and the jet cutting method. Nozzle resonance method (called prilling in chemistry) involves extruding a liquid through a nozzle as a jet and applying a vibration to the liquid (through a membrane) or the nozzle. This method will be largely described in Chapter 14. Very large production may be reached (tons a day) by using multinozzle systems. Size may range from a few micrometers to millimeters. However, for high viscous liquid, damping effects limit the possibilities to reduce the size lower than 800 µm. Commercial devices exist at pilot or even industrial scale (Inotech, Switzerland; Sodeva, France).

Cutting the jet with a little "wheel" has been proposed to overpass the limitation of the nozzle resonance method in regard to the liquid viscosity [Vorlop, 1994]. Inner liquid core of the microcapsules is extruded through a nozzle as a jet. The jet is cut by wire fixed on a rotating wheel (Figure 1a). The falling droplets are collected in solidifying bath (cold water, cross-linking agent solution). The system has been patented [Vorlop, 1994] and is commercialized by geniaLab, Germany (Figure 1c).

▦▦ Materials

Cutting jet method

The device includes:

- a high speed motor (1000 to 40 000 rpm, in function of the droplet size)

- a rotating wheel (Figure 1a and b)

- an injection system composed of
    - either vessel under pressure or pump without pulsations, and
    - connected to small injectors

- a collecting vessel with a gentle agitation.

The whole system may be incorporated in a close reactor to insure aseptic conditions.

A. home made apparatus (ENSAIA)

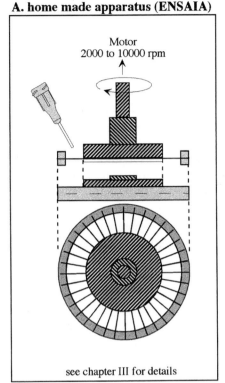

see chapter III for details

B. Lost evaluations

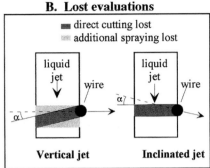

C. bottom vue genialab system

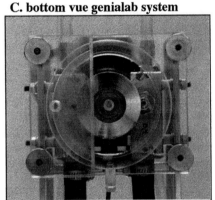

Fig. 1. Jet cutting method

Procedure

From a practical point of view, engineers will first fix the diameter of the droplets, d. The flow rate, F, may be computed as the products of the droplet volume times the frequency of cutting (rotational speed, ω time the number of wires, z):

Driving conditions

$$F = \frac{\pi}{6} d^3 (z\ \omega) \eta \tag{1}$$

where η is the yield (fraction of the extruded liquid recovers as microcapsules). The flow rate may be obviously multiplied by using several injectors.

To evaluate the yield, it is necessary to determine the fraction of the jet lost (ejected) during the cutting (see Pruesse, 1998a et 1998b for complete analysis). While the wire passes through the jet, the jet itself moves (Figure 1b). To minimise the cutting loss, the jet may be inclined in regard to the wheel (Figure 1b). The optimal jet-to-wheel angle, α for minimum cutting loss is given by (Pruesse, 1998 b):

$$\sin\ \alpha = \frac{u_{jet}}{U_{wires}} = F\frac{\frac{\pi}{4}d_i^2}{2\pi\ l\ \omega} = \frac{2\ F}{\pi^2 d_i^2 l\ \omega} \tag{2}$$

where u_{jet} and u_{wires} are respectively the linear velocity of the jet and wires, l the distance between the centre of wheel and the cutting position, d_i the jet diameter (assimilated to the internal nozzle diameter). The jet-to-wheel angle α must however be kept in a working range of 45 to 90 $^\circ$.

In optimum conditions, the loss is mainly a jet cylinder with a height equal to the wires diameter, d_w (Figure 1d). The fraction of liquid lost $(1-\eta)$ is mainly given by the ratio between the cutting lost cylinder and the jet cylinder between two cuts. This ratio may easily be assimilated to the ratio between wire diameter and the distance between wires.

$$1 - \eta = \frac{d_w z}{2\pi D} \tag{3}$$

The main constraint to reduce the lost fraction is the wire diameter. For 0.4 mm wire, lost fraction is ranging between 5 and 10 %. Using very fine wires (0.1 mm), it becomes negligible (less than 2 %). In practice, the loss may be slightly higher. geniaLab in particular observes the need to clean the wheel to remove material glued on it. This phenomena is however difficult to predict as it may vary strongly from material to material.

Results

The jet cutting method allows to produce beads or capsules ranging from 250 μm (3 l/h) to 2 mm (1500 l/h). The size distribution is relatively small (5 to 10 %) and the yield included between 90 and 100 %.

Subprotocol 2
Lentikats

Principle   Generally, microcapsules are relatively spherical. The Federal Agriculture Research Center of Braunschweig (FAL) has developed a large scale method based on lens-like shape microcapsule production (Figure 2a) [Jekel, 1998]. Droplets are extruded on a flat surface where they solidify in the form of lenses (either internal chemical or thermal gelation). The process is developed by geniaLab (Germany) at lab and industrial scales under the trademark LentiKats. Lens shape avoids mass transfer while providing good settling properties.

Procedure

Materials and Procedure

Figure 2a presents a general view of LentiKats production. Figure 2b shows LentiKats production device for pilot scale. Through a series of nozzles, droplets are extruded on a large moving strip. If required, spraying, heating or cooling system may be placed on top of the strip. At the end of the apparatus, solidified LentiKats are detached by the curvature of the strip (or with help of a cutter).

Results

LentiKats have usually 3 to 4 mm in diameter and 200 to 400 μm in thickness. Assuming 20 nozzles and a strip speed of 1 cm per second, the productivity is 20 kg per hour. Such a process implicates the use of encapsulating material of very good mechanical strength. Currently, LentiKats are made from polyvinyl alcohol (PVA) which shows very good mechanical stability. Several months of continuous stirring didn't cause a decrease in stability.

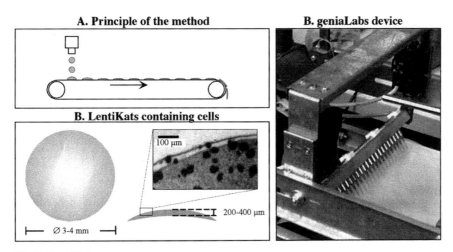

A. Principle of the method

B. geniaLabs device

B. LentiKats containing cells

100 μm

200-400 μm

Ø 3-4 mm

Fig. 2. LentiKat process

Subprotocol 3
Rotating Devices

Principle   While a liquid is flowed onto a rotating disk, it is ejected by centrifugation as droplets, ligaments or continuous film in function of the rotating speed, disk size, liquid flow rate and liquid properties. Film breaks into ligaments, ligaments into droplets. The flow rate is generally high (hundreds of litres per hour) and rotating devices are very promising for very large flow rates.

Unfortunately, the theories are quite complex and could not be summarized here. Moreover, due to missing information a complete description of the apparatus is not posssible. As an example, most authors do not give the droplet ejection distance. Most settings of the system are given as adimensional numbers which don't allow the return to real apparatus dimensions.

■ ■ Materials

In all cases, the system will comprise:

- an injection system composed of
  - a vessel under pressure or unpulsed pump
  - connected to an injector

- a rotating device which must avoid vibration as much as possible. Four devices have been proposed:
  - simple disk
  - rotating cup (Figure 3a) [Hinze, 1950]
  - vibrated disk (Figure 3b) [Chicheportiche, 1993]
  - nozzle fixed on a rotating cylinder (Figure 3c) [Schlameus, 1993]

- a collecting device containing the solidifying solution.

The complete system may be included in a closed chamber to insure sterility.

■ ■ Procedure

Running description   The simple disk allows limited control of the flow characteristics. The three other systems have then been proposed to improve the characteristic of the droplet formation. Rotating cup allows improvement of flow characteristics and reduction of the droplet ejection distance [Hinze, 1950]. In the

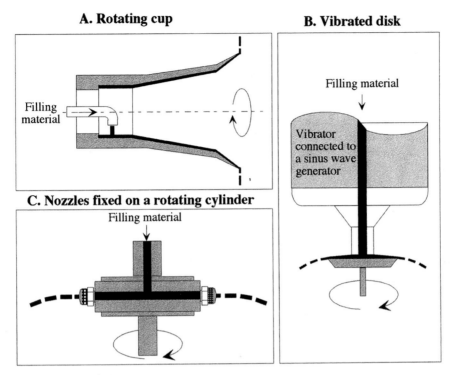

Fig. 3. Rotating devices

ligament formation state, controlled vibration may be applied to the disk and ligaments will break in quasi monodispersed droplets [Chicheportiche, 1993]. Although the theory is more complex, the principle is similar to the nozzle resonance method (Chapter 14). Finally, flowing the liquid through a nozzle on a rotating cylinder allows us to form a jet of controlled characteristics. It may be possible to combine such technology with the nozzle resonance technology [Schlameus, 1993].

Results

In all cases, the size distribution is narrower with rotating disk (standard deviation as low as 5 %) than with pressure atomisers (30 to 50 %). It allows us to reach a large scale (tons per hour). The ejection diameter may however be quite large (up to 10 m), limiting the square meter productivity and requiring very large collecting bath. In such conditions, sterility may be difficult to maintain. Vertical collectors have been proposed (Figure 3e) but no information is available to prove their efficiency.

Subprotocol 4
Emulsification Using Static Mixer

Principle   Several encapsulation technologies are based on the emulsification of an aqueous phase in an oil phase (see Chapter 3). Emulsification may be realized in turbine reactor (batch). However, at large scale, the production of microcapsules may be realized by passing both phases through a static mixer (Figure 4c). The advantages comprise continuous processing, small installation, easy sterilizing, low shear, narrower size distribution.

▪▪ Materials

Figure 4a schematises a complete device for producing microcapsules using static mixer. Figure 4b presents a detailed system including the injection and the static mixer. The material will then include:

- three tanks (dispersed phase, continuous oil phase, solidifying solution)

- three pumps (the process is relatively insensible to small pulses. Internal phase is often viscous, screw or moineau pump, PCM Pumps, France, allow to pump it)

- the emulsification device (Figure 4b) containing the static mixer (Sulzer, Switzerland) (Figure 4c)

- the collecting device (see below)

▪▪ Procedure

Regarding the building of the device the following advice may be given:

- Dispersed phase (microcapsule core) must be injected at the centre of the tube.

- Dispersed solution injector diameter will be designed to insure equal linear velocity of the dispersed and continuous phase (50 % of external tube for 25 % dispersed phase of total flow).

A. Principle of the method

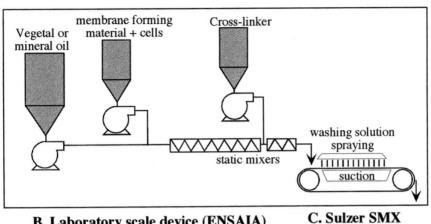

B. Laboratory scale device (ENSAIA)   C. Sulzer SMX

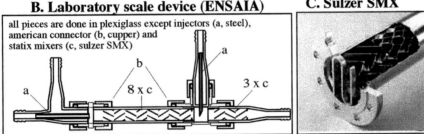

Fig. 4. Static mixer systems

- Selection of the static mixer depends on the solution properties, flow rate. Sulzer SMX mixers provide good emulsification with limiting shear capacity.

- Static mixer must not move in the tube but not be fixed to allow easy washing (in case of accidental solidification in the static mixer).

- Flow rate of the solidifying solution must be limited (maximum 10 % of the total flow rate).

- Tubing after homogenization may be prolonged (as spiral) to insure complete solidifying.

- Microcapsules may be collected in a tank and separated from oil later. A better solution would involve collecting microcapsules on a vibrating moving strip with bottom aspiration and top washing solution spraying.

From the point of view of the driving conditions, it must be pointed out that:

- Static mixers allow low droplet size (as low as 100 µm) to be attained without requiring emulsifier.

- Standard deviation of size distribution may be as low as 30 % in case of low viscosity dispersed phase (less than 100 mPas · s). It may be necessary to heat the solution (30 to 40°C).

- Dispersed phase must also be limited to maximum 25 to 30% of the total flow.

- The bead size is a function of many parameters. As a first approach, using 10 mPas · s dispersed phase and continuous phase composed of ethyl cocoate (20 mPas · s), a SMX static mixer (1 cm) was found whose size correlated to:

$$d = 45 \ u^{-1.2} \tag{4}$$

where d (mm) is the droplet diameter and u (m/s) the linear velocity in the static mixer. This law may be independent of the static mixer diameter. Other experiments show that the droplet size is mainly proportional to the inverse of the continuous phase viscosity.

- the residence time in the emulsifying and mixing zones is very small (0.1 to 0.5 s). To succeed to get nice capsules, it is then very important to realise the solidification step in a very short period. Assuming that the process of solidification it-self is fast, this may be achieved by:
  - introducing chemical cross-linkers soluble in continuous oil phase or
  - cooling the droplets directly by addition of cold oil after emulsification

Results

In the absence of emulsifiers, microcapsules ranging from 100 µm to 1 mm may be obtained. For 1 cm diameter SMX Sulzer static mixer, productivity ranges from 10 to 25 kg/h (respectively 250 to 600 kg/h for 5 cm mixer), allowing very large production.

References

Chicheportiche J.M. (1993) Etude de la Fragmentation commandee des Jets liquides issus d'un disque en rotation et réalisation d'un générateur de gouttelettes mono-dispersées. PhD thesis, University of Paris VI, May 5.

Hinze J.O., Milborn H. (1950) Atomisation of liquid by means of a rotating cup. J. Appl. Mechanics 145-153

Pruesse U., Fox B., et al. (1998a) New process (jet cutting method) for the production of spherical beads from highly viscous polymer solutions. Chem. Eng. Technol. 21: 29-33

Pruesse U., Bruske F., et al. (1998b) Improvement of the jet cutting method for the preparation of spherical particles, from viscous polymer solutions. Chem. Eng. Technol. 21: 153-157

Schlameus W. (1993) Centrifugal encapsulation. in: Risch S.J., Reineccius G.A. Encapsulation and controlled release of food ingredients, ACS Symposium series 590, Chapter 9, pp 96-103

Vorlop K.D., Breford J. (1994) German Patent N° DE 44249998, 15.07.1994

Suppliers

geniaLab BioTechnologie, Hamburger Str. 245, 38114 Braunschweig, Germany (Fax: 49-531 23 21 0 22)

Inotech AG, Kirchstrasse 1, 5605 Dottikon, Switzerland
(Fax: 41-56 624 29 88)

PCM pompes, BP 35, 92173 Vanves, France (Fax: 33-1-41 08 15 00)

Sodeva, BP 299, 73375 Le Bourget du Lac cedex, France
(Fax: 33-4-79 26 12 65)

Sulzer, 8401 Winterthur, Switzerland

Immobilization at Large Scale
with the Resonance Nozzle Technique

JAN H. HUNIK

Introduction

Immobilization for bioprocesses requires a diameter range of the gel-beads between 0.5 to 3 mm. Uniformly shaped gel-beads with a narrow diameter distribution are preferred in many applications. The resonance nozzle technique can provide this with relatively high flow rates. The resonance nozzle technique is based on the regular break-up of a liquid jet of gel-forming material. A gel-forming material such as κ-carrageenan is used as an example in this chapter. The objective of this chapter is to provide the engineering and technical background to produce regular gel droplets with a predefined diameter, a specific flow rate and a high yield.

The key event in making regular shaped gel-beads is the break-up of the liquid gel into droplets. The emphasis in this chapter will be on the calculations for large-scale production of regular gel droplets. The calculations can be applied for others gel materials mentioned in this chapter, when the appropriate liquid gel properties are measured or known.

Theory

A basic set of equations necessary to determine the experimental set-up is introduced here. The theoretical background of the equations used is extensively discussed in Hunik et al. (1993).

Jan H. Hunik, DSM Food Specialties, Wateringseweg 1 , Delft, 2600 MA, The Netherlands (*phone* +31-15-2793916; *fax* +31-15-2792490; *e-mail* Jan.Hunik@dsm-group.com)

Jet formation

Different regimes of liquid flow emerging from a small outlet can be distinguished. Drops are formed at the orifice at very low flow rates. A capillary jet with a laminar-flow profile will be formed at increasing flow rates. With an additional increase in flow rate a capillary jet with a turbulent-flow profile is formed. Eventually the liquid at the orifice will be atomized. We are interested in a jet with a laminar-flow regime. This flow regime is limited by a minimum flow velocity at which dripping changes to a capillary jet with a laminar-flow profile

$$u_j \geq 6.45 \sqrt{\frac{\sigma_l}{\rho_l \cdot d_t}} \tag{1}$$

With u_j as the flow velocity [m/s] calculated from the flow rate Q [m³/s] and the cross section surface area [m²] of the orifice. The diameter of the liquid jet d_j is smaller than the orifice diameter d_t. The relation between d_j and d_t is

$$d_j = \frac{\sqrt{3}}{2} \cdot d_t \tag{2}$$

Break-up of viscous-liquid capillary jets in air

Disturbances in the liquid flow cause constrictions of the jet and eventually the break-up in uniform droplets. For the formation of uniform droplets, the maximum flow velocity at which this uniform break-up regime can occur is

$$u_j < 22. \left[\frac{\mu_l^{0.28} \cdot \sigma_l^{0.36}}{\rho_l^{0.64} \cdot d_t^{0.64}} \right] \tag{3}$$

Above this liquid flow velocity the liquid jet will break-up into non-uniform droplets.

Relation between droplet size and break-up frequency

The droplet size d_p as a function of the applied frequency f and flow rate Q can be predicted with a mass balance. When we assume that from each sinus wave one droplet originates, the following mass balance can be set up

$$d_P = \left[\frac{Q \cdot 6}{f \cdot \pi} \right]^{1/3} \tag{4}$$

Optimal wavelength

The frequencies and corresponding wavelengths for breaking up the capillary jet must be within a certain interval. For wavelengths considerably shorter or longer than the jet diameter it is less probable that break-up of the jet will occur.

The optimal wavelength λopt for breaking up the capillary jet is given by Weber (1931):

$$\lambda_{opt} = 4.44 \cdot d_j \left[1 + \frac{3\mu_l}{\sqrt{\rho_l \cdot \sigma_l \cdot d_j}} \right] \tag{5}$$

The frequency f and wavelength λ are related by the liquid flow velocity according to

$$f = \frac{u_j}{\lambda_j} \tag{6}$$

With eqs (5) and (6), the optimal frequency $fopt$ can be calculated.

Procedure

The entrapped cells are produced by first mixing a cell suspension with a liquid gel solution. The suspension passes the resonance nozzle so that droplets are formed, which are collected in a stirred solution for hardening. Then the produced beads are transferred to the medium where cell growth can start.

Gel material

Immobilization methods for living cells require high cell viability after immobilization. Common immobilization materials are alginate and carrageenan. These materials are processed as liquid gel. The concentration of gel material is 2.6 % and 2-3 % (w/v) for, respectively, a κ-carrageenan or alginate gel. It is advised to make the liquid gel in cold water with a homogenizer. Add the total amount of alginate or carrageenan powder to the water when stirred vigorously by the homogenizer. Gelling of the powder takes a few seconds after it has been in contact with water. With this rapid dispersion in cold water the powder particles are fully dispersed before the rapid increase in viscosity starts.

The liquid gel mixed with the cell suspension is transferred to a temperature controlled vessel that can be pressurized. A sight-glass in this vessel is recommended. A pressure valve and an (optional) magnetic flowmeter control the outflow of the vessel to the resonance nozzle.

A solid gel material is not formed before the droplets are in a hardening solution. This hardening solution contains 0.2 M CaCl$_2$ or 0.75 M KCl ions for respectively, alginate and κ-carrageenan gel. Problems with coalescence of the droplets at the surface of the hardening solution can be avoided by the use of an organic solvent layer, like n-decane on top of the gelling liquid (Buitelaar et al., 1988). The organic solvent will also retain the spherical shape of the gel beads. A non-toxic solvent should be used and tested before use in the immobilization. Low concentrations of Ca^{2+} or K$^+$ should be present in the final medium used for the biocatalyst that is immobilized. The ions of Ca^{2+} or K$^+$ for, respectively, alginate and κ-carrageenan gel-beads are essential to maintain a solid gel.

Resonance-nozzle apparatus

In the extrusion technique the aqueous gel solution is pressed at such a high flow rate through a small orifice that a jet is formed. With a membrane, a sinusoidal vibration of a certain frequency is transferred to the liquid. This signal will cause the break-up of the jet into uniform droplets.

In Figure 1 the principle of the resonance nozzle apparatus is shown. The sinusoidal pulse for the vibration is generated by a sinus generator. This signal is amplified and transferred to a vibration exciter. With this vibration exciter the signal is transferred to a flow through chamber where the vibration is transferred to the liquid. The break-up of the liquid jet will take place after the orifice. Droplet formation can only be observed with the stroboscope. A schematic design of the flow through chamber is shown in Figure 2.

Some elements in the design are crucial for the final result. The membrane should be as rigid as possible. With a rigid metal plate as membrane and only a small edge of flexible rubber between the metal plate and the housing, disturbances of the vibration are avoided.

The actual orifice should be very smooth without any roughness. The only disturbance of the jet is the applied vibration. Other disturbances will interfere and lead to irregular break-up. Drilled holes in a metal plate are not smooth enough, because the edges are, on micro scale, frayed by the drilling equipment. We applied precious stones, i.e. ruby, as used in watches. These precious stones have a very smooth hole in the centre,

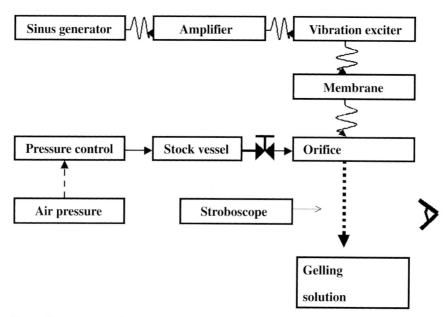

Fig. 1. Resonance nozzle set-up

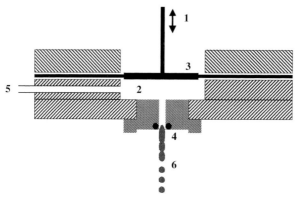

Fig. 2. Resonance nozzle with connection to the vibration exciter (1) and membrane (3), the chamber (2), where the vibration is transferred to the liquid. The liquid enters the apparatus at (5) and is extruded at the orifice (4). Break-up will occur shortly after the orifice (6)

for the original application as bearing for the cogwheels. We obtained these rubies with various "orifice" diameters at a jeweller's shop.

Depending on the apparatus used the maximum production can be limited by the maximum power transfer of the vibration exciter to the liquid.

Procedure to obtain a specific diameter of the gel-beads

Shrinkage of droplets in the gelling solution

First the density of the gel solution is determined. Before immobilization the desired gel bead diameter has to be defined. The gel material will shrink in the hardening process and this shrinkage has to be taken into account.

For the determination of the shrinkage, approximately 200 gel-beads are produced with the dripping method (Hulst et al. 1985). These droplets are collected in the appropriate gelling solution for hardening. Volume of the collected droplets is determined by the increase in weight of the gelling solution. The number and diameter of the gel beads is measured and the shrinkage (based on volume) is calculated from these data.

Viscosity

The theory for the break-up of capillary jets is based on liquids with a Newtonian viscous behaviour. The κ-carrageenan solution, for example, is a non-Newtonian fluid. For non-Newtonian fluids the viscosity depends on the shear rate. Thus the appropriate viscosity of the liquid gel solution should be determined experimentally. As a valid approach for the equations presented in this chapter the highest dynamic viscosity (μ_l) as measured at shear rates below 0.1 [1/s] can be used. This can be measured in commercially available viscometers.

Calculations

1. Obtain the relevant liquid properties. Experimental or from literature. The density in [kg/m^3], surface tension σ in [N/m] and the dynamic viscosity μ in [Pa · s]. The values for, for example water, are a $\rho = 1000$ kg/m^3, s = 0.072 N/m (without surface active components) and μ_l = 0.001 Pa · s (at 20°C).

2. For most of the gelling materials the viscosity has a non-Newtonian behaviour. This implies that the appropriate viscosity should be used as obtained from the viscosity versus shear rate measurement of the liquid gel.

The calculations after step 3 have an iterative character. It is advised to use a spreadsheet program for the calculations. The equations presented are valid for a gel droplet d_p range of 0.5 to 3.0 [mm] and viscosity range μ of 0.001 to 1 [Pa · s] covering most of the commonly used gel materials.

3. Start with a desired flow rate Q per orifice and use [m³/s] as the unit for the calculations. With [m³/s] you will get low values for the flow rate as 1 litre/hour is equivalent to 2.8 *10⁻⁷ m³/s. Also the orifice diameter d_t [m] should be set. It is advised to start with an orifice diameter d_t that $0.3 < d_p > 1$.

4. Calculate the cross section surface area [m²] of d_t.

5. Calculate the u_j with: Q / (outcome of Step 4). Check if the free falling velocity is larger than u_j. In Figure 3 the free falling velocity of liquid droplets with density ρ = 1000 [kg/m³] is shown. If this free falling velocity is smaller than u_j return to step 3 and adjust flow rate Q or orifice diameter d_t.

6. Calculate the lower limit for u_j with equation (1). Calculate the upper limit for u_j with equation (3). If u_j is between these boundaries then continue with step.

7. Otherwise return to step 3 and adjust the flow rate Q.

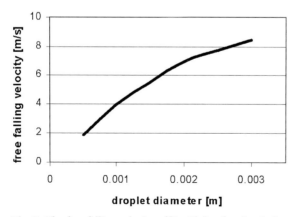

Fig. 3. The free falling velocity of liquid droplets in air. Based on a density ρ of 1000 kg/m³

7. Calculate d_j with equation (2).

8. Calculate the optimal wavelength λ_{opt} for breaking up the capillary jet with equation (5) and the optimal frequency f_{opt} with equation (6). Check the frequency range of the function generator to see whether this frequency is applicable.

9. Calculate the droplet diameter d_p with equation (4) When the droplet size is outside the desired range return to step 3 and adjust flow rate Q or orifice diameter d_t and repeat the schedule.

Example

In the example the production of κ-carrageenan gel-beads is demonstrated with the resonance nozzle apparatus.

Preparation of the gel solution

The κ-carrageenan (Genugel 0909) powder was obtained from the Copenhagen Pectin Factory. The aqueous κ-carrageenan solution was prepared with cold tap water. The κ-carrageenan (2.6 % w/w) was rapidly dissolved with a Silverson Homogenizer. The temperature was increased to 35 °C and the liquid gel was stored in a temperature-controlled vessel before using it in the resonance nozzle apparatus. The immobilization was done at a temperature of 35 °C, at which temperature most cells will survive. Based on experience, the viscosity at 35 °C was low enough for handling the liquid gel in the resonance nozzle apparatus.

Resonance-nozzle apparatus

The resonance nozzle apparatus used in this example was equipped with 6 nozzles. Vibration transfer was done with a large and rigid metal plate connected to the vibration exciter and 6 membranes. A Tandar TG 102 function generator generates the sinusoidal pulse for the vibration. This signal is amplified by a Bruel & Kjaer, type 2706, amplifier with a maximum power consumption of 75 W. The amplified signal is transferred to a Bruel & Kjaer vibration exciter, type 4809, in which the signal is transferred to a vibration of the membrane in the six vibration chambers of the nozzle. In the experiments orifices with a diameter of 0.6 mm or 0.8 mm were used. The frequency was varied between 50 and 600 s^{-1}. The produced droplets were visualized with a Griffin Xenon Stroboscope type 60.

Shrinkage of droplets in the gelling solution

A density ρ of 1008 [kg/m³] for the 2.6 % (w/w) κ-carrageenan solution was determined. The shrinkage of the droplets was 6.9 % (volume) after hardening in the gelling solution and transfer to the final medium.

Viscosity measurements

The viscosity of the κ-carrageenan solutions was measured with an Ostwald viscosity meter and a Haake Rotovisco RV 20 rotation viscosity meter at 35 °C.

The viscosity of the 2.6 % (w/w) κ-carrageenan solution at 35 °C as a function of the shear rate is shown in Figure 4. A value of 0.73 [Pa · s] was used as dynamic viscosity μ in the calculations.

Calculations

The relevant liquid properties were: density ρ = 1008 [kg/m³], surface tension σ = 0.072 [N/m] and dynamic viscosity μ = 0.73 [Pa · s].

For droplets of 2 mm a free-falling velocity of 7.2 m/s was obtained from Figure 3. This is above the maximum flow rate in experiments described here. Merging of successive droplets was thus not likely to happen and indeed not observed.

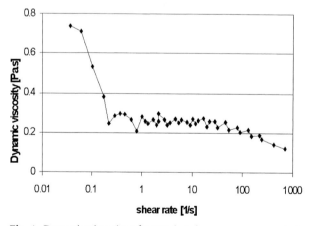

Fig. 4. Dynamic viscosity of 2.6 % (w/w) κ-carrageenan solution as a function of the shear rate

The theoretical lower and upper limit for u_j were calculated. A lower limit of 2.25 and 1.89 m/s is found for, respectively, the 0.8- and 0.6-mm orifices. For the theoretical maximum velocity of the jet, values of 8.97 and 10.7 m/s were calculated for, respectively, the 0.8- and 0.6-mm orifices. The maximum velocity in our experiments was 2.97 and 4.11 m/s for, respectively, the 0.8- and 0.6-mm orifices. The maximum flow rate in the experiments did not exceed the limits described above, but was limited by the power transfer of the vibration exciter to the liquid. The power output of the apparatus at the highest flow rates was the maximum for the amplifier used. However, the theoretical limitations should be taken into account when other combinations of liquid, orifice and vibration exciter are used.

The visual observations, together with the measured gel-droplet diameters are presented in Table 1. For pictures of the gel-droplets formed with the resonance nozzle, see Hunik et al. (1993).

The visual observations showed that not all the jets were nicely broken up into uniform droplets. At vibration frequencies above the optimal frequency we observed the formation of a chain of droplets connected by thin threads of the liquid. The formation of droplet chains is also observed by Goldin et al. (1969), who studied the stability of capillary jets for several visco-elastic fluids. The thin threads between the droplets seem to be more stable than the jet itself. This chain with droplets and threads is also found in the hardening solution.

Besides chain formation satellites are also formed at vibration frequencies below the optimal frequency. According to Rutland & Jameson (1970), satellites are always formed when a capillary jet breaks up into droplets. A minimum of satellites should be formed around the optimum frequency (f_{opt}), which corresponds well with our results. The theoretical droplet diameters in Table 1 correspond well or were slightly smaller compared to the measured droplet size, and the assumption of one droplet for each wavelength seems valid over a broad range of flow rates and frequencies.

The optimal break-up frequency (f_{opt}) was calculated with the assumed dynamic viscosity of the jet, $0.73 = [Pa \cdot s]$. This optimal frequency matched well with the experimental results. The standard deviation of the bead diameter was the lowest around the optimal frequencies (Table 1).

The maximum flow rate where uniform droplets were produced with this six-point resonance nozzle apparatus was 27.6 $dm^3 \cdot h^{-1}$ with a bead diameter of 2.13 mm and a standard deviation of 0.12 mm. After hardening and shrinkage, gel beads with a diameter of 2.08 mm were obtained.

Table 1. Visual observations of the liquid-jet break up with the 0.8 and 0.6 mm orifice.

Flow rate [dm³/h] Q	[m/s] uj	Fopt eq (6) [1/s]	Frequency applied [1/s]							
			50	100	150	200	250	300	400	600
						with 0.8 mm orifice				
20.5	1.89	187			+	2.25 0.07 2.07	-	-	-	
23.8	2.19	217		++	+	2.56 0.29 2.19	2.47 0.36 2.02	2.36 0.38 1.92	-	
27.6	2.54	252			+	2.43 0.23 2.29	2.13 0.12 2.13	-	-	
32.2	2.97	294		++	+	+	2.94 0.56 2.24			
						with 0.6 mm orifice				
14.8	2.42	300	++	+	2.47 0.17 2.07	2.39 0.29 1.86		1.92 0.31 1.64	-	
17.0	2.78	344	++	+	2.30 0.18 2.17	1.97 0.11 1.96		1.71 0.12 1.71	1.30 0.53 1.55	
19.5	3.19	395	++	+	2.75 0.38 2.26	2.34 0.19 2.04		1.77 0.16 1.79	-	
25.1	4.11	508	++	+	+	2.38 0.32 2.23		2.02 0.37 1.95	1.75 0.25 1.77	-

Visual observations: – – = no break up, – = no break up and chain formation, + = break up in droplets of irregular shape, ++ = break up with satellite droplets.

Numerical values Top: Average diameter [mm]

Middle: standard deviation [mm]

Bottom: theoretical diameter (eq 4) [mm]

References

Buitelaar RM; Hulst AC; Tramper J (1988) Immobilization of biocatalysts in thermogels using the resonance nozzle for rapid drop formation and an organic solvent for gelling. Biotechnol Techniques 2:109-114.

Goldin M; Yerushalmi J; Pfeffer R; Shinnar R (1969) Break-up of a laminar capillary jet of a viscoelastic fluid. J Fluid Mech 38:689-711.

Hulst AC; Tramper J; Riet K van 't; Westerbeek JMM (1985) A new technique for the production of immobilized biocatalyst in large quantities. Biotechnol Bioeng 27:870-876.

Hunik JH; Tramper J (1993) Large-scale production of κ-carrageenan droplets for gel-bead production: theoretical and practical limitations of size and production rate. Biotechnol. Prog. 9:186-192

Rutland DF; Jameson GJ (1970) Theoretical prediction of the sizes of drops formed in the break-up of capillary jets. Chem Eng Sci 25:1689-1698.

Weber C (1931) Zum zerfall eines Flüssigkeitsstrahles. Ztschr fur angew Math und Mech 11:136-154.

Wijffels RH; Tramper J (1989) Performance of growing Nitrosomonas europaea immobilized in κ-carrageenan. Appl Microbiol Biotechnol 32:108-112.

Abbreviations

d	diameter [m]
f	frequency [s^{-1}]
Q	flow rate [m^3/s]
u	velocity [m/s]
λ	wavelength [m]
μ	dynamic viscosity [Pa · s]
ρ	density [kg/m]
σ	surface tension [N/m]
j	jet
l	liquid
p	droplet
t	orifice
opt	optimal

External Mass Transfer

René H. Wijffels

Introduction

Numerous papers have been devoted to modelling simultaneous diffusion and conversion of substrate by immobilized biocatalysts. In such systems the substrate is transferred from a liquid phase to a solid phase in which the reaction occurs. In many of these studies it was assumed that the rate-limiting step in the process is the diffusion of substrate through the solid phase, because the thickness of the boundary layer surrounding the solid particles was supposed to be much smaller than the radius of the gel bead used as a support for the immobilized biocatalyst. Furthermore, the diffusion coefficient in the carrier is usually smaller than in water. In the case of immobilized-cell processes, however, external mass transfer needs to be incorporated in the model in order to obtain realistic predictions (De Gooijer et al. 1991, Wijffels et al. 1991). With immobilized growing bacteria a relatively thin layer of biomass will be formed just beneath the support surface. If for example oxygen is the limiting substrate, this active layer will be in the order of 100 µm thick, which is of the same order of magnitude as the thickness of the boundary layer.

Relatively little is known, however, about external mass transfer in bioreactors as usually the relative velocity of the particles is not known. The relations used are based on empirical equations and need to be adopted not only for all types of bioreactors used, but also for all scales and procedures. In practice new correlations should be derived based on experiments for all processes.

René H. Wijffels, WageningenUniversity, Food and Bioprocess Engineering Group, P.O. Box 8129, Wageningen, 6700 EV, The Netherlands (*phone* +31-0317-484372; *fax* +31-0317-482237; *e-mail* rene.wijffels@algemeen.pk.wag-ur.nl)

In this chapter it is described how to determine k_{il} in air-lift loop re-actors as a function of superficial gas velocity. It is, however, not possible to determine k_{il} directly with immobilized-cell systems, since the resistance to mass transfer is not only determined by external-transport limitations. In the case of ion-exchange resins absorption rates are so high that the exchange rate is determined entirely by external mass transfer. Therefore ion-exchange resins are used as a solid phase and from the results relations between the Sherwood number and superficial gas velocity are derived.

Outline

In Chapter 9 calculation of diffusion limitation has been described. For calculation of the external effectiveness factor the exernal mass transfer coefficient was necessary. The external mass transfer coefficient is dependent on turbulance in the liquid medium. In this chapter a method is described to determine the external mass transfer coefficient. Apart from that relations for the external mass transfer coefficient can be obtained from literature. In this chapter a number of such relations will be given.

Theory

Liquid/solid mass transfer descriptions are often based on the film theory. This means, the liquid/bulk phase is assumed to be perfectly mixed except for a thin boundary layer surrounding the particles. It is assumed that mass transfer across this layer solely occurs by diffusion.

Mass transfer across the stagnant layer can thus be described by:

$$F_i'' = \frac{D_{il}}{\delta}(C_{sb} - C_{si}) \quad (mol \cdot m^{-2} \cdot s^{-1})$$

As δ is generally not known, a liquid/solid mass-transfer coefficient, k_{il}, which can be calculated with the dimensionless Sherwood number, is defined as:

$$k_{il} = \frac{D_{il}}{\delta} = \frac{Sh D_{il}}{d_p} \quad (m \cdot s^{-1})$$

Correlations for the Sherwood number are partly theoretical and partly empirical. For a sphere surrounded by an infinitely extended stagnant medium it can be derived that Sh equals 2. If the liquid starts to flow, the value of Sh increases. In that situation the stagnant region is limited to the near surroundings of the bead. The thickness of that stagnant layer, δ, decreases as the flow increases, while k_{il} and Sh increase. The general form of the Sherwood equation is:

$$Sh = 2 + aSc^\beta \, Re^\gamma$$

The dependence of the Sherwood number on the Schmidt and Reynolds numbers is usually described by empirical correlations.

In the book Basic Bioreactor Design (van 't Riet and Tramper 1991) the liquid/solid mass-transfer coefficient (k_{il}) is calculated with the relation of Ranz and Marshall (1952) or Brian and Hales (1962) (Table 1). It is assumed that the particles move with the rate of free fall. For this the Reynolds number was calculated from the Galileo number. At lower Reynolds numbers (Re < 30) the relation of Brian and Hales (1962) can be used, at Reynolds numbers larger than 30, the relation of Ranz and Marshall (1952) is applied (Table 1). These relations are used in stirred tank reactors.

Whether this assumption is appropriate for application in air-driven bioreactors such as bubble columns and air-lift loop reactors is not without doubt. Because of the gas supplied and the presence of a large solid phase fraction, particles will be disturbed in their free fall velocity. This means that the particle movement is influenced by the turbulent flow in the reactor and is hindered because of the high particle density.

Another approach which is frequently used is Kolmogoroff's theory of local isotropical turbulence. Turbulence in air-driven reactors is caused by

Table 1. Sherwood relations used for liquid/solid mass transfer in bubble columns.

Sano et al. (1974)	$Sh = 2 + 0.4 \, Sc^{0.33} \left(\frac{\varepsilon_e d_p^4}{v_l^3} \right)^{0.25}$
Sänger and Deckwer (1981)	$Sh = 2 + 0.545 \, Sc^{0.33} \left(\frac{\varepsilon_e d_p^4}{v_l^3} \right)^{0.264}$
Livingston and Chase (1990)	$Sh = 2 + 0.275 \, Sc^{0.33} \left(\frac{\varepsilon_e d_p^4}{v_l^3} \right)^{0.274}$
Kamawura and Sasano (1965)	$Sh = 2 + 0.72 \, Sc^{0.33} \left(\frac{\varepsilon_e d_p^4}{v_l^3} \right)^{0.208}$
Wijffels et al. (1997)	$Sh = 2 + 0.56 \, Sc^{0.33} \left(\frac{\varepsilon_e d_p^4}{v_l^3} \right)^{0.2}$
Ranz and Marshall (1952)	$Sh = 2 + 0.6 \, Sc^{0.33} \left(\frac{\varepsilon_e d_p^4}{v_l^3} \right)^{0.5}$
Brian and Hales (1969)	$Sh = \sqrt{4 + 1.21 \, (ScRe_p)^{0.67}}$

the introduction of air. Introduced air bubbles will rise and as they rise expand (Chisti 1989). This isothermal expansion of the gas bubbles is the main source of power input in the reactor. The rising bubbles will generate large eddies, which in first instance cause large-scale mixing in the reactor. Fluid packages will be interchanged. Further mixing will occur because the initial large eddies scatter into smaller ones. Those larger eddies transfer their kinetic energy to the smaller ones which will scatter again and so on. This transfer of energy occurs in all directions and as eddy-generation proceeds the orientation of the eddies will be lost and turbulence therefore becomes isotropic. Ultimately the only information transferred from the large eddies to the very small ones is the energy which will finally be dissipated (by friction to heat) in the smallest eddies having a flow which is just laminar (Re=1). On the basis of a dimension analysis the size of these smallest eddies can be derived and from that a definition of the Reynolds number of a spherical particle in an isotropic turbulent flow field (Kawase and Moo-Young 1990, Shah et al. 1982, Deckwer 1980, Shinnar and Church 1960):

$$Re = c\left(\frac{\varepsilon_e d_p^4}{v_l^3}\right)^{1/3} \quad for \quad d_p \ll \lambda$$

The energy dissipation rate in a bubble column can be calculated by Chisti (1989):

$$\varepsilon_e = g v_{gs}$$

For an air-lift loop reactor a correction is made for the part of the reactor that is not aerated:

$$\varepsilon_e = \frac{g v_{gs}}{1 + \frac{A_d}{A_r}}$$

Although the theoretical basis for application of Kolmogoroff's theory in airlift-loop reactors is not without doubt, implementation of $\varepsilon.d_p^4/v_l^3$ in the Sherwood number has been successful in correlating liquid-solid mass transfer data. The difference between the so defined Reynolds number and the Reynolds number as used in the equations of Ranz and Marshall and Brian and Hales is that it is dependent on the applied energy to the reactor and not on the density difference between solid and liquid phase.

Materials

- HCl
- NaOH
- air-lift loop reactor of 3.4 dm^3 with diameters of riser and downcomer 65 and 45 mm, respectively
- Brooks instruments, mass flow 5850 TR
- pH meter
- agar
- petri-dish
- Amberlite-IRA 458 (acrylic)
- Amberlite-IRA 996
- image analyser

Procedure

Experiments were executed in air-lift loop reactors with an external loop. As a solid phase anionic-exchange resins were used. The rate of absorption of chloride ions by the resins after addition of a HCl-solution was measured by monitoring the rate of decrease of the concentration of cotransported H$^+$-ions (Appendix). This rate is determined by external mass transfer (Appendix). Diffusion of the Cl$^-$ ions across the main part of the boundary is only influenced by H$^+$-ions and constant diffusion coefficient over the entire stagnant layer can be assumed (Appendix).

1. Reactor is filled with water and a known amount of particles.

2. Two types of ion-exchange resins are used in this case; wetted Amberlite-IRA 458 (acrylic) and Amberlite-IRA 996 with a total exchange capacity of 1250 mol · m^{-3} both. The resin concentration in the reactor was 0.82 and 0.68 % (v/v) respectively.

3. For determination of the particle-size distribution, 100-300 particles are mixed with 10 cm^3 of an agar solution (1.2 % (w/v)) and poured in a petri-dish (diameter 9 cm) (Tramper et al. 1984). After gelation photographs were taken. The sauter diameter of particles was determined by means of image analysis with a Magiscan 2.

4. Measure the density of the particles as has been described in Chapter 4. The density of carrageenan is also measured in order to apply the relation for e.g. carrageenan beads.

5. Airflow is controlled by a mass-flow controller (Brooks instruments, mass flow 5850 TR). Experiments were done at different superficial gas velocities ranging from 0.17 to 1.25 cm $\cdot$ s^{-1}.

6. The pH value was measured continuously in the downcomer.

7. Control temperature in the reactor to a constant value (in our case 30 °C).

8. Regenerate the ion-exchange resin in the reactor with 20 dm^3 of a NaOH solution (23.5 mmol. dm^{-3}) before every experiment. The NaOH solution is pumped through the reactor with a flow of 5 cm^3 s^{-1}. The excess regenerant is replaced by pumping 25-30 dm^3 of demineralized water through the reactor until a pH value of 7 is reached. Also during regeneration air is injected in order to suspend the particles.

9. After regeneration 3.5 cm^3 of a HCl solution (2.4 mol. dm^{-3}) is added. The pH in the liquid phase at the start of such an experiment is then 2.5.

10. Measure the absorption rate of submitted Cl$^-$ to the ion-exchange particles by measuring the corresponding change in pH as a function of time.

Results

The sauter diameter of the Amberlite-IRA 996 and Amberlite-IRA 458 was 1.19 and 0.80 mm, respectively. The density of Amberlite-IRA 458, Amberlite-IRA 996 and carrageenan were 1080, 1130 and 1008 kg $\cdot$ m^{-3}, respectively.

At several superficial gas velocities pH versus time curves were determined. After regeneration of the ion-exchange resins HCl was added. During the first minutes the pH increased linearly with time. It was assumed that the concentration of Cl$^-$ at the surface of the resins was 0 and that the Cl$^-$ and H$^+$-ions were codiffusing over the entire boundary layer (appendix). In that case k_{il} can be calculated from the initial slope of the pH-time curves by means of:

$$\ln \frac{C_{sb(0)}}{C_{sb(1t)}} = k_{il}at$$

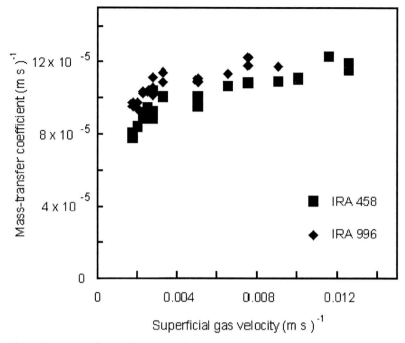

Fig. 1. Mass-transfer coefficient as a function of the superficial gas velocity for Amberlite-IRA 458 and Amberlite-IRA 996

The results of these measurements are given in Figure 1, both for Amberlite-IRA 458 and Amberlite A-IRA 996 as a function of the superficial gas velocity. It is shown that there is an effect of the superficial gas velocity on the value of k_{il}, in particular at the lower gas velocities. The results can be correlated to Sherwood relations as shown in Figure 2. The Sherwood number based on Kolmogoroff's theory was used. Double log plots of the data show straight lines from which Sherwood relations can be derived by linear regression (Wijffels et al. 1998):

$$Sh = 2 + 0.56 \ Sc^{0.33} \left(\frac{\varepsilon_e d_p^4}{v_l^3} \right)^{0.2}$$

Other Sherwood relations obtained from literature for liquid/solid mass transfer in bubble columns are given in Table 1.

Sherwood numbers for κ-carrageenan gel beads suspended in air-lift loop reactors were calculated with all equations given in Table 1. Experimental conditions from a previous study were used (Wijffels et al. 1991).

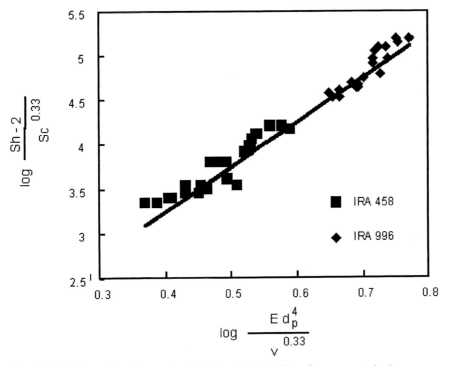

Fig. 2. Double-log plots of experimental data with fitted line (linear regression)

This means, the cross-sectional area of the riser and downcomer were 0.00332 and 0.00159 m², respectively, and the diameter of the carrageenan beads 2.08 mm. Sherwood numbers were calculated at different superficial gas velocities. The results are shown in Figure 3. All relations show an increase of the Sherwood number with increasing superficial gas velocity except for the relation of Brian and Hales (1952) as expected. In the case of the relation of Brian and Hales it was assumed that particles move with the rate of free fall and therefore the driving force is the density difference between the liquid and solid phase. In this case Brian and Hales was used instead of Ranz and Marshall (1969) as the Reynolds number of the moving particles was below 30.

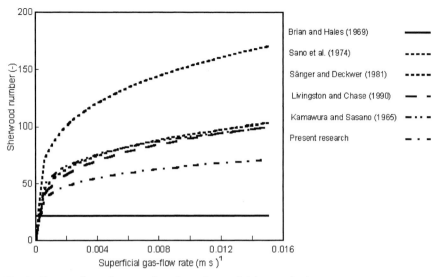

Fig. 3. Sherwood number as a function of superficial gas velocity

Troubleshooting

One additional effect should be accounted for as well, i.e. the particle hold-up. It is well known that the particle hold-up influences the hydrodynamics considerably. Verlaan (1987) showed that a particle hold-up of 25% gave a reduction of 50% in the k_la for gas/liquid mass transfer. A similar effect on liquid/solid mass transfer seems to be reasonable. This means, that further research should be done at particle hold-ups that are equal to the hold-ups in processes with immobilized biocatalysts (15-25%).

References

Brian P.L.T. , Hales H.B., Effects of transpiration and changing diameter on heat and mass transfer to spheres. AIChE Journal, 14 (1969) 419-425.

Chisti M.Y., Airlift bioreactors. Elsevier Applied Science, London, (1989) 345 p.

Deckwer W.D., Physical transport phenomena in bubble column bioreactors-II. liquid-solid mass transfer, heating and heat transfer. In: eds: M. Moo-Young, C.W. Robinson , C. Vezina, Advances in Biotechnology. Proc. 6th Int. Ferment. Symposium London, Pergamon Press, Toronto 1 (1980) 471-476.

De Gooijer C.D., Wijffels R.H., Tramper J. (1991). Growth and substrate consumption of *Nitrobacter agilis* cells immobilized in carrageenan. Part 1: Dynamic modeling. Biotechnol. Bioeng. 38: 224-231

Helfferich, F., Ionenaustauscher. Band I: Grundlagen. Verlag Chemie, Weinheim, Germany, (1959) 520 p.

Helfferich, F., Ion- exchange kinetics. V. Ion exchange accompanied by reactions. J. Phys. Chem., 69 (1965) 1178-1187

Kawase Y., Moo-Young M., Mathematical models for design of bioreactors: application of Kolmogoroff's theory of isotropic turbulence. Chem. Eng. J., 43 (1990) B19-B41.

Ranz W.E. , Marshall W.R., Evaporation from drops, part II. Chem. Eng. Progr., 48 (1952) 173-180.

Livingston A.G., Chase H.A., Liquid-solid mass transfer in a three phase draft tube fluidized bed reactor. Chem. Eng. Comm., 92 (1990) B19-B41

Rodrigues, A.E., Dynamics of ion exchange processes. In: L. Liberti, F.G. Helfferich (eds.). Mass transfer and kinetics of ion exchange. Martinus Nijhoff Publicers, The Hague, the Netherlands, (1983) 259-311

Sano Y., Yamaguchi N., Adachi T., Mass transfer coefficients for suspended particles in agitated vessels and bubble columns. J. Chem. Eng. Japan 7 (4) (1974) 255-261

Sänger P., Deckwer W.D., Liquid-solid mass transfer in aerated suspensions. Chem. Eng. J., 22 (1981) 179-186.

Shah Y.T., Kelkar B.G., Godbole S.P. , Deckwer W.D., Design parameters estimations for bubble column reactors. AIChE Journal, 28 (1982) 353-379.

Shinnar R., Church J.M., Statistical theories of turbulence in predicting particle size in agitated dispersions. Ind. Eng. Chem., 52 (1960) 253-256.

Tramper, J., Van Groenestijn, J.W., Luyben, K.Ch.A.M. , Hulshoff Pol, L.W., Some physical and kinetic properties of granular anaerobic sludge. In: E.H. Houwink , R.R. van der Meer (eds.), Innovations in Biotechnology, Elsevier Science Publishers B.V., Amsterdam, The Netherlands (1984) 145-155

Van 't Riet K., Tramper L. (1991) Basic Bioreactor Design. Marcel Dekker Inc.

Verlaan P., Modelling and characterization of an airlift-loop reactor. PhD thesis, Wageningen Agricultural University, The Netherlands (1987).

Wijffels R.H., De Gooijer C.D., Schepers A.W., Beuling E.E., Mallée L.R., Tramper J. (1995) Growth of immobilized *Nitrosomonas europaea*: implementation of diffusion limitation over microcolonies. Enzyme and Microbial Technology 17: 462-471

Wijffels R.H., Verheul M., Beverloo W.A., Tramper J. (1998) Liquid/solid mass transfer in an air-lift loop reactor with a dispersed solid phase. J. Chem. Technol. Biotechnol. 71: 147-154.

Abbreviations

a	liquid/solid specific surface area (m^{-1})
A_d	downcomer cross sectional area (m^2)
A_r	riser cross sectional area (m^2)
c	constant (-)
d_p	particle diameter (m)
D_{il}	diffusion coefficient of i in liquid phase ($m^2 \cdot s^{-1}$)
$D_{e,g}$	diffusion coefficient in solid phase ($m^2 \cdot s^{-1}$)
g	gravitational acceleration ($m \cdot s^{-2}$)
F_i''	flux ($mol \cdot m^{-2} \cdot s^{-1}$)
k_{il}	liquid-solid mass transfer coefficient for component i ($m \cdot s^{-1}$)

N	relative half life times of film transport and gel transport (-)
R	gas constant (J $\cdot$ mol^{-1} $\cdot$ K^{-1})
Re_p	particle Reynolds number (-)
C_i	ion concentration (mol $\cdot$ m^{-3})
C_{sb}	substrate concentration in bulk liquid phase (mol $\cdot$ m^{-3})
C_{si}	substrate concentration solid-liquid interphase (mol $\cdot$ m^{-3})
F	constant of Faraday (C $\cdot$ mol^{-1})
Sc	Schmidt number (-)
Sh	Sherwood number (-)
t	time (s)
u	velocity (m $\cdot$ s^{-1})
u_g	mean superficial gas velocity in riser (m $\cdot$ s^{-1})
V	electrochemical potential (V)
z_i	valence of ion (-)
v_{pl}	particle velocity relative to the liquid (m $\cdot$ s^{-1})
v_{gs}	mean superficial gas velocity (m $\cdot$ s^{-1})
α	constant (-)
β	constant (-)
γ	constant (-)
δ	thickness of stagnant layer (m)
ϵ_e	specific energy dissipation per unit of mass (m^2 $\cdot$ s^{-3})
λ	length scale of microscale eddy (m)
η_l	dynamic viscosity (kg $\cdot$ m^{-1} $\cdot$ s^{-1})
v_l	kinematic viscosity (m^2 $\cdot$ s^{-1})
ρ_l	density liquid phase (kg $\cdot$ m^{-3})
ρ_p	density particles (kg $\cdot$ m^{-3})
η	0.7975 x 10^{-3} kg $\cdot$ m^{-1} $\cdot$ s^{-1} [20]
v	0.801 x 10^{-6} m^2 $\cdot$ s^{-1} [20]
ρ_l	995.65 kg $\cdot$ m^{-3} [20]
ρ_{IRA458}	1080 kg $\cdot$ m^{-3}
ρ_{IRA996}	1130 kg $\cdot$ m^{-3}
ρ_{carr}	1008 kg $\cdot$ m^{-3}
$d_{23-1458}$	0.80 mm
$d_{23-1996}$	1.19 mm
D_{O2}	2.8 x 10^{-9} m^2 $\cdot$ s^{-1} [21]
D_H	10.06 x 10^{-9} m^2 $\cdot$ s^{-1} [20]
D_{Cl}	2.28 x 10^{-9} m^2 $\cdot$ s^{-1} [20]

Determination of mass-transfer coefficients with ion-exchange resins

The ion-exchange resins used were anionic. During the experiment diffusion of H$^+$ and Cl$^-$ across the boundary layer will occur; the OH$^-$ ions will be exchanged for CL$^-$ ions and a neutralisation reaction takes place.

As ions are carriers of an electrical charge their transport will be influenced by electrical forces and Fick's law of diffusion is not valid. Therefore the relation of Nernst-Planck is used (1959, 1965):

$$F_i'' = -D_i(\Delta C_i + z_i \frac{F}{RT} \Delta V)$$

Ions generally diffuse in two directions. The movement of the ions is influenced by all ions present and for all ions separate Nernst-Planck equations can be derived. During ion-exchange, there is always electro-neutrality and there is no electric current. The equation above thus reduces to:

$$F_i'' = -D_i \Delta C_i$$

In our case there are three ions involved, i.e. OH⁻, Cl⁻ and H⁺:

$$C_{OH^-} + C_{cr} = C_{H^+}$$

The difference between D_i in the equation given above and the equation of Fick is that in the given equation D_i is not constant as its value depends on the concentration of all ions. As the ion concentration in the boundary layer is not constant, D_i is also not.

In this case the exchange reaction is combined with a neutralisation reaction, however, this diffusion coefficient will be constant. At low pH,

$$C_{OH^-} \ll C_{H^+}$$

If the concentration of OH⁻ can be neglected:

$$C_{Cl^-} = C_{H^+}$$

D_i will then be constant and can be described by Helfferich et al. 1959, 1965) and Rodrigues (1983):

$$D_i = \frac{2D_{H^+} D_{Cl^-}}{D_{H^+} + D_{Cl^-}}$$

This equation is only valid in case the OH⁻ concentration is negligible across the entire boundary layer. During the experiments mass-transfer coefficients were determined by measuring the pH increase as a function of time after addition of HCl to suspended ion-exchange resins. The pH of well-rinsed, regenerated particles will be 7. After addition of HCl the pH in suspension was 2.5, which makes the equation applicable. Across such a boundary layer, however, an ion-concentration gradient will occur. Near the surface of the bead the concentration of OH⁻ may be much higher. In the worst case the pH at the resin surface will be 7. To what extent this will influence the local diffusion coefficients, can be simulated. Concentrations of H⁺ and OH⁻ are calculated as a function of the distance from the particle surface.In the bulk phase the concentration H⁺ was 10^{-3} mol·m⁻³ and at the liquid/solid interface 10^{-9} mol·m⁻³. A linear concentration profile for H⁺ across the stagnant layer was assumed. Only in a minor part of the stagnant layer is the concentration of OH⁻ higher than H⁺. At a relative distance of 0.001d C_{OH-} is 100 times smaller than C_{H+} and it may be assumed that diffusion in this case is determined by H⁺ and Cl⁻.

Another assumption which was made, was that mass transfer across the stagnant liquid layer was the rate-limiting process during ion exchange. This was done by comparing half life times of film diffusion and particle diffusion (Helfferich 1959, 1965):

$$N = \frac{12 X D_{e,g} \delta}{C_i D_{li} \cdot d_p}$$

If N is much larger than 1, the rate will be determined by film diffusion. In our experiments N ≫ 1 as X was 1 x 10³ eq m⁻³. Mass transfer is thus entirely determined by external mass transfer.

Dimensionless numbers

Sherwood : $Sh = \dfrac{k_{il} d_p}{D_{il}}$

Reynold : $Re_p = \dfrac{\rho_p v_{pl} d_p}{\eta_l}$

Schmidt : $Sc = \dfrac{v_l}{D_{il}}$

Galileo : $Ga = \dfrac{g d_p^3 \rho_l (\rho_p - \rho_l)}{\eta_l^2}$

$Ga < 36$ then $Re_p = \dfrac{Ga}{18}$

$36 < Ga < 8^* 10^4$ then $Re_p = 0.153\, Ga^{0.71}$

$8^* 10^4 < Ga < 3^* 10^9$ then $Re_p = 1.74\, Ga^{0.5}$

Liquid Fluidization of Gel-Bead Particles

ERIK VAN ZESSEN, JOHANNES TRAMPER, and ARJEN RINZEMA

Introduction

Fluidization of particles occurs when the velocity of a liquid flowing through a packed bed of particles becomes larger than a certain threshold value, which is called the minimal fluidization velocity. At the minimal fluidization velocity the pressure drop (ΔP) over a packed bed of particles equals the specific buoyant weight of the particles:

$$\Delta P = V_p(\rho_p - \rho)g \tag{1}$$

V_p is the volume of the particles per unit area of the bed, ρ_p is the density of the particles, ρ is the density of the liquid and g is the gravity constant.

When the liquid velocity is increased beyond the minimal fluidization velocity the bed expands. The particles are not supported by neighbouring particles but by the up flowing liquid; the height of the bed increases. As the height increases, the total bed volume increases naturally, and the volume of particles per total bed volume (i.e. the solids hold-up) decreases.

When the liquid velocity is increased up to the minimal fluidization velocity, the pressure drop per unit height of bed increases. When this velocity is increased beyond the minimal fluidization velocity the pressure drop per unit height of bed decreases, as the overall pressure drop is constant at fluidization and equal to the specific buoyant weight of the particles.

✉ Erik van Zessen, Wageningen University, Food and Bioprocess Engineering Group, Biotechnion PO Box 8129, Wageningen, 6700 EV, The Netherlands
(*phone* +31-0317-4-82240; *fax* +31-0317-4-82237;
e-mail Erik.vanZessen@algemeen.pk.wau.nl)
Johannes Tramper
Arjen Rinzema

ΔP/L (Pa/m)

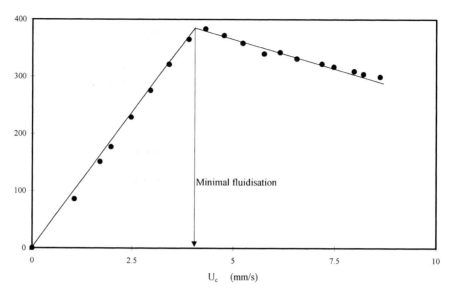

Minimal fluidisation

U$_c$ (mm/s)

Fig. 1. Pressure drop per unit bed height as function of superficial liquid velocity. Gel-bead particles with diameter of 2.8 mm, and density of 1060 kg/m³. Liquid is 30 mM KCl, temperature is 25° C

The liquid velocity can be increased until the settling velocity of a single particle is reached. These above-mentioned observations are universal for solid particles, so they also apply to gel-bead particles. A typical example of pressure drop per unit bed height of gel-bead particles for liquid velocities going beyond the minimal fluidization velocity is shown in Figure 1. An excellent review on liquid fluidization has recently been published by Di Felice (1995). Gòdia and Solà (1995) give a review on fluidized-bed bioreactors and a list of applications of those bioreactors.

A commonly accepted model to predict the voidage of a fluidized bed is the empirical relation of Richardson and Zaki (1954):

$$U_c = v_{\varepsilon \to 1} \varepsilon^n \tag{2}$$

U_c is the superficial velocity of the liquid, $v_{\varepsilon \to 1}$ is the superficial velocity at a voidage equal to one, and n is an empirical constant.

The solids hold-up (ε_s) follows directly from the fact that the sum of both fractions equals 1.

$$\varepsilon_s = 1 - \varepsilon \tag{3}$$

Models for n as well as models for v_∞ are abundant in literature. Hartman et al. (1992) give an overview of different models for n. The $v_{\varepsilon\to1}$ is not always equal to the settling velocity of a single particle, for a detailed discussion see Di Felice (1995). If $v_{\varepsilon\to1}$ is equal to the terminal settling velocity, different models from literature for predicting the terminal settling velocity can be used. However, as will be shown below, these models cannot be applied to predict the solids hold-up of gel-bead particles. Consequently the parameters n and v_∞ have to be determined experimentally.

So far liquid fluidization of gel beads has been mentioned. When working with immobilized micro-organism, direct air supply might be needed. Introducing gas in a liquid fluidized bed automatically creates a gas-liquid-solid three-phase fluidized bed. Muroyama, and Fan (1985) give a good review on engineering aspects of this three-phase fluidized bed. This review does not incorporate gel-bead particles. Engineering aspects on such three-phase fluidized beds using gel beads can be found in Verlaan and Tramper (1987). The remaining part of this chapter deals with two-phase liquid fluidization of gel beads.

Materials

A typical experimental set-up for liquid fluidization of gel beads looks as simple as depicted in Figure 2. A column of transparent material, e.g. glass, perspex etc. is the core of the set-up. Liquid is pumped with pump (1) from the storage vessel through a calibrated flow meter (2) to the column. Use a pump with a constant flow rate, e.g. a centrifugal or gear pump. The first part of the column is filled with glass beads (3) to provide a good flow distribution over the gel beads that make up the second part of the column (5). A sieve plate with a relative large spacing (4) is used to support the packed bed of gel beads. To control temperature both column and storage vessel can be jacketed.

Experimental set-up

Procedure

First the column is filled with the fluidizing liquid and gel beads with a known total volume are added (the volume can be determined as described in the chapter "Measurement of Density, Particle Size and Shape of Support"). Next the flow rate is increased and the bed height is recorded. Flow rate is measured with a calibrated flow meter and, as the liquid fluidized bed shows a quiescent behaviour, bed height can be measured with a ruler,

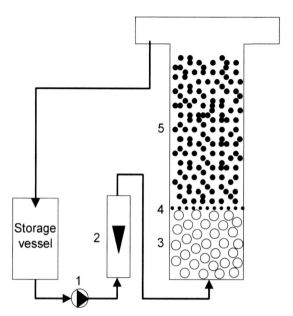

Fig. 2. Experimental set-up for liquid fluidization. 1 pump, 2 flow meter, 3 glass bead (o), 4 sieve plate, 5 gel beads (•)

or any equivalent instrument. Make sure that the bed height is constant after a flow-rate increment.

Results

As an example of liquid fluidization of gel beads, results are shown and discussed for the fluidization of alginate beads. Gel beads were made from an 8% alginate solution mixed with yeast cells (25 v%) by the dripping method. This resulted in alginate beads with a density of 1039.9 kg/m^3, a Sauter mean diameter of 4.27 mm and a circularity of 1.00. These beads were stored in a 50 mM CaCl$_2$ solution. An experimental set-up as described above was used for the fluidization experiments with calibrated rotameters as a flow meter. The column was 1.5 m in height and had an inner diameter of 6 cm. The temperature was equal to the ambient temperature, i.e. 25°C.

Figure 3 shows the bed height as a function of the superficial liquid velocity (50 mM CaCl$_2$); bed height increases more than linearly with the superficial liquid velocity. The arrow indicates the point of minimal fluidization.

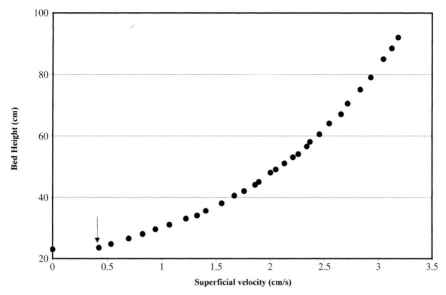

Fig. 3. Bed height of alginate gel beads as function of the superficial 50 mM $CaCl_2$ solution's velocity

The measured bed height can be converted to the solids hold-up with equation (4):

$$\varepsilon_s = \frac{V_{\text{beads}}}{0.25 \ \pi \ D^2_{\text{column}} \ H_{\text{bed}}}$$

Figure 4 shows the solids hold-up as a function of the superficial velocity. The measured data were fitted with equation 2 and 3, resulting in a $v_{\varepsilon \to 1}$ of 4.67 cm/s and a parameter n of 2.10. This fit describes the data well as shown in Figure 4. So solids hold-up, and hence bed height, can be accurately predicted with equations 2 - 4 and the fitted parameters.

The $v_{\varepsilon \to 1}$ is not equal to the settling velocity of a single alginate gel bead, which has been measured in the same column and is equal to 5.22 cm/s. A detailed discussion about this discrepancy is beyond the scope of this manual; for more information see Di Felice (1994). Literature models used for calculating the parameter n in equation 2 resulted in larger values, e.g. the model of Richardson and Zaki (1954) resulted in 2.53. This higher value for parameter n will result in a lower solids hold-up, see Figure 4. A full discussion on the fact that the fitted parameter n is smaller than predicted with literature models is beyond the scope of this manual. We did not

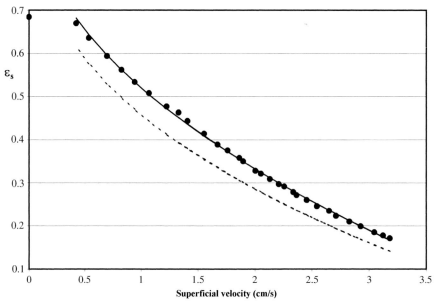

Fig. 4. Solids hold-up of alginate beads as a function of the superficial 50 mM $CaCl_2$ solution's velocity. data, model with fitted constants $v_{\varepsilon\to 1}$=4.67; n=2.10, model with $v_{\varepsilon\to 1}$ = 4.67; n=2.53

find experimental data on the solids hold-up for liquid fluidization of gel beads in the literature.

This example of liquid fluidization of alginate beads shows that equation 2 and 3 can be used to fit experimental data satisfactorily, but the parameters used cannot be predicted with established literature models. Consequently these parameters have to be determined experimentally.

Acknowledgements. The Association of Biotechnology Centers in the Netherlands (ABON) is acknowledged for their financial support, and A. Mulder is acknowledged for the supply of alginate gel beads.

References

Di Felice, R., Chem. Eng. Sci. , 1995, 50(8) 1213 -1245

Foscolo P.U., Gibilaro L.G., Waldram S.P., Chem. Eng. Sci. 38(8) 1983 1251 - 1260

Gòdia F. and Solà C., Biotechnol. Prog, 1995, 11, 479 - 497

Muroyama K., Fan L.S., 1985, AICHE J., 31, 1-34

Richardson J.F. and Zaki W.N., 1954, Trans. Instn. Chem. Engrs., 32, 35 - 53

Hartman M., Havlin V., Svodoba K., Chem. Eng. Sci., 1989, 44(11), 2770 - 2775

Turton, R. and Clark, N.N., 1987, Powder Technol. 53, 127-129

Verlaan P., Tramper J., Hydrodynamics axial dispersion and gas-liquid oxygen transfer in an airlift-loop bioreactor with three-phase flow. In: Bioreactors and Biotransformations (Moody G.W., Baker P.B., eds.) Elsevier, Barking, UK, 1987, 363-373

Zigrang, D.J.,and Silvester, N.D., 1981, A.I.Ch.E J. 27, 1043-1044

Gradients in Liquid, Gas or Solid Fractions

René H. Wijffels

Introduction

Processes consist of different steps. If for example an aerobic conversion takes place in a solid particle the oxygen has to be transferred from the gas phase to the liquid phase and from the liquid phase to the solid phase. The rate of the overall process will be determined by the slowest process. Optimization of the process can be achieved by optimization of the slowest process step. In that case, however, an overview has to be made of all possible process steps. A useful tool for such an inventory is regime analysis. A system with immobilized biocatalysts has a complex behaviour. A complete description of the process for a wide range of conditions is time consuming or even impossible. This argument is valid for most biotechnological processes and a consistent approach to simplify these processes is regime analysis (Moser 1988, Roels 1983). In the regime analysis presented by Schouten et al. (1986), with immobilized *Clostridium spp.* for isopropanol/butanol production, the effectiveness factor for the immobilized cells was estimated to be 1. They conclude that the isopropanol/butanol production is not diffusion controlled and the immobilized cells behave as free cells. New for the regime analysis presented here is the addition of a solid third phase with immobilized cells growing in a diffusion-controlled situation.

Regime analysis has been worked out by Hunik et al. (1994). For this the nitrification process with immobilized cells was worked out at optimal temperatures and at low temperatures with low substrate concentrations and at inhibitory substrate concentrations.

René H. Wijffels, Wageningen Agricultural University, Food and Bioprocess Engineering Group, P.O. Box 8128, Wageningen, 6700 EV, The Netherlands (*phone* +31-0317-484372; *fax* +31-0317-482237; *e-mail* rene.wijffels@algemeen.pk.wag-ur.nl)

Large-scale applications of this immobilized-cell process are limited to a few plants because this complicated process is difficult to scale up (Heijnen et al. 1991). A better understanding of the rate-limiting factors, important for scaling up, can be obtained with a regime analysis (Sweere et al. 1987). From such a regime analysis a set of design directives for the different applications of immobilized biocatalysts can be derived.

Regime analysis can either be used for the optimization of the reactor design or to reveal the rate-limiting step of a process, see Figure 1 (Sweere et al. 1987). Optimization of the reactor design requires several iterations until a previously defined optimum is obtained. Here we are interested in the rate-limiting step of the process and not in optimization of the reactor design. The regime analysis starts with an inventory of all transport and conversion mechanisms of the process. The characteristic time of each mechanism is then estimated; relatively slow mechanisms have a high characteristic time, while lower characteristic times apply to faster mechanisms. The comparison of characteristic times for conversion and transport mechanisms of a particular substrate can thus reveal the rate-limiting step.

Air-lift loop reactors are most suitable for immobilized-cell processes. They lack mechanical stirring and are easy to scale up. Mechanical stirring

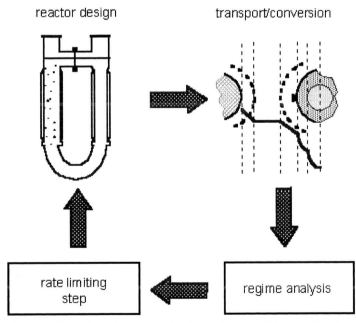

Fig. 1. Optimization of a process by determination of the rate-limiting step with regime analysis

can cause abrasion of the immobilization material and should therefore be avoided. Air-lift loop reactors are characterized, among others, by Verlaan (1987) and Chisti (1989) with respect to their liquid circulation, mixing properties and mass transfer.

Outline

This chapter contains only simple calculations to determine the rate limiting step and to determine whether gradients in liquid, gas or solid fractions can be expected. Calculations are worked out for a nitrification process with immobilized cells. The type of bioreactor used is an air-lift loop reactor. The same calculations can be used for other organisms and reactor types as there are only differences in kinetic parameters and values of mass transfer coefficients. In the 'References' some books are listed that give relations for mass transfer for other reactor types as well.

Procedure

Regime analysis

For nitrification with cells non-homogeneously growing in a gel bead several transport mechanisms for the substrates (O_2 and NH_4^+), intermediate (NO_2^-) and product (NO_3^-) can be distinguished: mass transfer of oxygen from air bubbles to the liquid phase; mass transfer of oxygen and ammonia from the bulk phase to the gel beads; mass transfer of oxygen, ammonia and nitrite within the beads to the cells, and mass transfer of nitrite (intermediate) and nitrate (product) from the beads to the liquid. All these transport mechanisms are characterized with either a mass-transfer coefficient or a diffusion coefficient. The conversion is characterized by the substrate consumption rate. This chain of transport and conversion should be combined with the reactor characteristics in the actual regime analysis. Mixing, circulation, gas and liquid retention times should be compared with the characteristic times for transport and conversion to reveal if gradients in the reactor bulk phase can be expected.

For this three types of characteristic times are defined:

1. Characteristic times of mass transfer.

2. Characteristic times of conversion.

3. Design specific characteristic times.

The design specific characteristic times determine whether gradients exist in the liquid or gas phase.

Characteristic times

Examples of relations for characteristic times are found in the literature (De Gooijer et al. 1991, Moser 1986, Sweere et al. 1987). In general these relations are obtained from the ratio between a capacity and a flow

$$\tau = \frac{capacity}{flow} \tag{1}$$

Capacity is defined as the available substrate for transport or conversion at process conditions. The flow is the rate of that particular transport or conversion. The characteristic time for mixing and circulation, which already have time as dimension, are directly used in the regime analysis.

When regime analysis is used for a three-phase system one phase must be used as a sort of "pivot" for the characteristic-time calculations. Most convenient is the continuous (liquid) phase, which is therefore, unless mentioned otherwise, used as "pivot" phase for the characteristic-time definitions given below.

Characteristic times of mass transfer

Oxygen is transported from gas phase via the liquid phase to the solid phase. The rate of this transport is determined by the characteristic times for gas-liquid mass transfer ($\tau_{lg}{}^{O2}$), liquid-solid mass transfer ($\tau_{ls}{}^{O2}$) and oxygen conversion ($\tau_{kin}{}^{O2}$).

The characteristic time ($\tau_{lg}{}^{O2}$) for the mass transfer of oxygen from the gas phase to the liquid phase is given by

$$\tau_{lg}{}^{O2} = \frac{1}{k_{lg} \cdot a_{lg}} \tag{2}$$

$$a_{lg} = a_g \cdot \frac{\varepsilon_g}{1 - \varepsilon_g}$$

$$a_g = \frac{6}{d_b} \tag{4}$$

The characteristic time (τ_{ls}) for the mass transfer from the liquid to the solid phase is given by

$$\tau_{ls} = \frac{l}{k_{ls} \cdot a_{ls}} \tag{5}$$

$$a_{ls} = a_s \cdot \frac{\varepsilon_s}{1 - \varepsilon_s} \tag{6}$$

$$a_s = \frac{6}{d_p} \tag{7}$$

The characteristic times for all the transport mechanisms in the reactor are independent of the substrate concentrations.

If a process is controlled by gas/liquid mass transfer of oxygen the design of the process can be improved with respect to oxygen transport, in particular between gas and liquid phase. This oxygen transport is dependent of k_{lg}, a_{lg}, k_{ls} and a_{ls}. Smaller gas bubbles and gel beads increase these values of k_{lg}, a_{lg}, k_{ls} and a_{ls}. The relation between a_{ls} and gel bead diameter (d_p) is shown in Eqs (6) and (7). A decrease in gel bead diameter (d_p) would increase the value a_{ls}. Nevertheless, the minimum gel bead diameter (d_p) is limited by the requirement to keep the gel beads in the reactor. The gas bubble diameter (d_b) is determined by the sparger and the coalescence behaviour of the medium, which is difficult to manipulate. For this reason it is not useful to produce small bubbles in a coalescent medium because the advantage of the small bubbles disappears shortly after the sparger. An increase in gas hold up (ε_g) would be beneficial for the oxygen-transfer rate according to Eq. (2), (3) and (4). An increase of solid phase hold up (ε_s), i.e. more gel beads in the reactor, is limited to a maximum of 35 %, and the k_{lg} will be reduced (Verlaan 1987).

Characteristic times of conversion

In contrast to the characteristic time of mass transfer, the characteristic time for substrate conversion (τ_{kin}) depends on the surface concentration of substrate (C_{si}) of the gel bead.

The characteristic time τ_{kin} for the substrate conversion by free cells is derived from the biomass concentration C_x (kg $\cdot$ m^{-3}), the substrate concentration C_s (mol $\cdot$ m^{-3}), and the kinetic parameters K_S (mol $\cdot$ m^{-3}), Y_{xs} (kg $\cdot$ mol^{-1}), and m_{max} (s^{-1}) of the relevant microorganism (Roels 1983):

$$\tau_{kin} = \frac{C_s}{\left(\frac{\mu_{max} \cdot C_x}{Y_{xs}}\right) \cdot \left(\frac{C_s}{K_S + C_s}\right)} \tag{8}$$

When we consider solid gel beads with immobilized cells, incorporate internal diffusion limitation and take into account the solid hold-up, equation 8 will reduce to:

$$\tau_{kin} = \left(\frac{Y_{xs}(K_s + C_{si})}{\varepsilon_s \cdot \eta_{ei} \cdot \mu_{max} \cdot C_{xi}}\right) \tag{9}$$

The concentration of the substrate at the interface (C_{si}) needs to be calculated from the flux = conversion relation given in chapter Chapter 15

$$k_{ls} \frac{6\varepsilon_s}{d_p} (C_{sb} - C_{si}) = \eta_{ei} \varepsilon_s \frac{\mu_{max} C_{si}}{K_s + C_{si} Y_{xs0}} \tag{10}$$

Design specific characteristic times

Transport of oxygen from the gas bubble to the liquid phase depletes the gas bubbles of oxygen. The characteristic time for the gas-bubble oxygen exhaustion (τ^O_{ex}) is given by

$$\tau^0_{ex} = \frac{L}{v_g} \tag{11}$$

The τ^O_{ex} in Eq. (11) is based on the gas phase, because the depletion of this phase is considered here.

The maximum gas-phase-retention time ($\tau_{ret}{}^{gas}$) in an air-lift loop reactor is calculated from the ratio of the reactor height and terminal rising velocity (approximately 0.25 m/s (Heijnen and Van 't Riet 1984)) of the gas bubbles:

$$\tau^0_{ret} = \frac{L}{v_g} \tag{12}$$

The actual value for gas-phase retention time ($\tau_{ret}{}^{gas}$) will be shorter when liquid circulation in the loop reactor is taken into account.

Oxygen depletion of the gas phase depends on the mass transfer of oxygen from gas to liquid phase and amount of oxygen available in the gas phase. When $\tau_{ex}{}^{O2}$ is smaller than $\tau_{ret}{}^{gas}$ we can conclude that complete exhaustion of the gas bubble is not likely to happen. In case it is just slightly smaller a significant decrease in oxygen concentration in the gas bubble will occur. As a result the characteristic time for oxygen gas/liquid transfer will be underestimated (τ_{lg}^{O2}). Oxygen exhaustion of gas bubbles in air-lift loop reactors does not become important below a reactor height of 15 m, but it also depends on the liquid velocity in the riser.

Characteristic times for liquid circulation, mixing and gas-phase retention time in air-lift loop reactors are related to the size of the reactor. The mixing time (τ_{mix}) for an air-lift loop reactor is calculated from the circulation time (τ_{circ}) as shown by Verlaan (1987):

$$\tau_{mix} = (4 \text{ to } 7) \cdot \tau_{circ} \tag{13}$$

This value of τ_{circ} can be measured easily in an existing reactor or calculated for a given reactor design (Verlaan 1987).

The values for the gas-phase hold up (ε_g) and liquid mixing time (τ_{mix}) depend on the reactor configuration and size. Gas hold up (ε_g) values for air-lift loop reactors will be in the range of 0.02-0.05 (Chisti 1989, Verlaan 1987). Van 't Riet and Tramper (1991) present some data about circulation and mixing times in air-lift loop reactors based on the model of Verlaan (1987). The mixing (180-600 s) and circulation times (45-150 s) are for two external air-lift loop reactors with a volume of 250 and 1225 m^3, respectively, a height to riser diameter ratio of 10 : 1 and a ratio between the riser to downcomer diameter of 2 : 1. A difference between mixing and circulation time should be made for loop reactors. Mixing times are useful for pulse-wise addition of a substrate to the reactor and circulation times are more useful for continuous addition of substrates in loop reactors. This latter situation is more applicable and circulation times were therefore used to predict gradients in the reactors.

The liquid-retention time (τ_{ret}^{liq}) is the reciprocal value of the dilution rate:

$$\tau_{ret}^{liq} = \frac{1}{D} \tag{14}$$

If the liquid circulation time (τ_{circ}) is in the same order of magnitude as the characteristic times for gas-liquid mass transfer (τ_{lg}^{O2}) and liquid-solid mass transfer (τ_{ls}^{O2}). Gradients in O_2 concentration in the liquid phase can be expected. A gel bead, when circulating through the loop reactor, will encounter rapid changes in O_2 concentrations as a consequence of these gradients.

The liquid phase can be assumed well mixed with if the liquid retention time (τ_{ret}^{liq}) is large compared to the circulation time (τ_{circ}).

Sufficient liquid mixing, e.g. prevention of oxygen gradients, should be provided for aerobic processes with immobilized cells. Oxygen gradients in the reactor can be decreased when the retention time in the non-aerated downcomer is short and the overall circulation time is kept as small as possible. Reactors should therefore have a relatively high riser to downcomer diameter ratio and small height to diameter ratio.

References

Chisti MY (1989) Airlift bioreactors, Elsevier science publishers, Essex, England. Heijnen JJ, Mulder A, Weltevrede R, Hols J, Leeuwen HLJM van (1991) Large scale anaerobic-aerobic treatment of complex industrial waste water using biofilm reactors. Wat Sci Tech, 23:1427-1436.

Chisti MY (1989) Airlift bioreactors, Elsevier science publishers, Essex, England Heijnen JJ, Riet K van 't (1984) Mass transfer, mixing and heat transfer phenomena in low viscostiy bubble column reactors. Chem Eng J 28:B21-B42

Heijnen JJ, Riet K van 't (1984) Mass transfer, mixing and heat transfer phenomena in low viscostiy bubble column reactors. Chem Eng J 28:B21-B42.

De Gooijer CD, Wijffels RH, Tramper J (1991) Growth and substrate consumption of *Nitrobacter agilis* cells immobilized in carrageenan: Part 1. dynamic modeling. Biotech Bioeng 38:224-231

Hunik J.H., Tramper J., Wijffels R.H. (1994a) A strategy to scale-up nitrification processes with immobilized nitrifying cells. Bioprocess Eng. 11: 73-82

Hunik JH, Bos CG, Hoogen MP van den, Gooijer CD de, Tramper J. (1994b) Validation of a dynamic model for substrate conversion and growth of *Nitrosomonas europaea* and *Nitrobacter agilis* cells immobilized in k-carrageenan beads. Biotechnology and Bioengineering 43: 1153-1163

Moser A (1988) Bioprocess technology: kinetics and reactors, Springer-Verlag, New York, USA.

Roels JA (1983) Energetics and kinetics in biotechnology, Elsevier Biomedical Press, Amsterdam, The Netherlands

Schouten GH, Guit RP, Zieleman GJ, Luyben KChAM, Kossen NWF (1986) A comparative study of a fluidized bed reactor and a gas lift loop reactor for the IBE process: Part 1. reactor design and scale down approach. J Cem Tech Biotechnol 36:335-343.

Sweere APJ, Luyben KChAM, Kossen NWF (1987) Regime analysis and scale-down: tools to investigate the performance of bioreactors. Enzyme Microb Technol 9:386-398.

Van 't Riet K, Tramper J (1991) Basic Bioreactor Design. Marcel Dekker Inc.

Verlaan P (1987) Modelling and characterization of an airlift-loop reactor, PhD Thesis, Wageningen Agricultural University, The Netherlands

Abbreviations

a_g	specific surface area of a gas bubble ($m^2 \cdot m^{-3}$ gas phase)
a_{lg}	surface area of the liquid/gas inter phase ($m^2 \cdot m^{-3}$ liquid phase)
a_{ls}	surface area of the solid/liquid inter phase ($m^2 \cdot m^{-3}$ liquid phase)
a_s	specific surface area of a gel bead ($m^2 \cdot m^{-3}$ solid phase)
C_s^*	saturation concentration (of oxygen) in the liquid phase ($mol \cdot m^{-3}$)
C_{si}	substrate concentration at surface of the biocatalyst ($mol \cdot m^{-3}$)
C_{sb}	substrate concentration in bulk phase ($mol \cdot m^{-3}$)
C_x	biomass concentration ($kg \cdot m^{-3}$)
d_b	gas bubble diameter (m)
d_p	biocatalyst particle diameter (m)
D	dilution rate (s^{-1})
H	Henry coefficient ($m^3 \cdot m^{-3}$)
K_s	half-rate constant ($mol \cdot m^{-3}$)
k_{ls}	liquid-solid mass transfer coefficient ($m \cdot s^{-1}$)
k_{lg}	gas-liquid mass transfer coefficient ($m \cdot s^{-1}$)
L	length of the column (m)
Y_{xs}	molar substrate yield ($mol \cdot kg^{-1}$)
v_g	terminal rising velocity of a gas bubble ($m \cdot s^{-1}$)
ε_g	gas hold up (m^3 gas $\cdot$ m^{-3} liquid)
ε_s	solid phase hold up (m^3 solid $\cdot$ m^{-3} liquid)
τ	characteristic time (s)
τ_g	τ for growth (s)
τ_{ex}^O	τ for oxygen exhaustion of gas bubbles (s)
τ_{lg}^O	τ for oxygen transfer from gas to liquid phase (s)
τ_{ret}^{liq}	τ for the liquid retention time (s)
τ_{ret}^{gas}	τ for the gas retention time (s)
τ_{ls}^i	τ for substrate (i) transfer from liquid to solid phase (s)
τ_{mix}	τ for mixing of liquid phase (s)
τ_{circ}	τ for liquid circulation in air-lift loop reactor (s)
τ_{kin}^i	τ for substrate (i) conversion (s)
τ_{conv}^i	τ for substrate (i) conversion in the biocatalyst (s)
μ_{max}	maximum specific growth rate (s^{-1})
η_{ei}	internal effectiveness factor (-)

Support Material Stability at the Process Conditions Used

EMILY J.T.M. LEENEN

Introduction

A support material should meet certain criteria to be useful for a specific application. The objective of this chapter is to characterize the stability of a certain support material at certain process conditions. In this way selection of the most suitable support material can be made. The selection of a suitable support material for wastewater treatment systems is described in more detail by Leenen et al. (1996).

The characteristics important for support material are:

1. solubility

2. biodegradability

3. mechanical stability

4. diffusivity

5. growth of entrapped cells

6. immobilization procedure

7. attachment of heterotrophic organisms

8. cost

In this chapter only the analysis of the stability of the support will be described (1-3). The measurement of the diffusivity is described in chapters Chapter 6 and Chapter 9, growth in chapters Chapter 8, Chapter 11 and

Emily J.T.M. Leenen, National Institute of Public Health and the Environment, Microbiological Laboratory for Health Protection, P.O.Box 1, Bilthoven, 3720 BA, The Netherlands (*phone* +31-302473711; *fax* +31-302744434; *e-mail* Imke.Leenen@rivm.nl or Frank-Imke@hetnet.nl.)

Chapter 12, and immobilization procedures in chapters Chapter 3, Chapter 13 and Chapter 14. As attachment of heterotrophic organisms and the costs are not described in this lab manual a few comments on these subjects are made at the end of this chapter.

Solubility The solubility of a support material can be investigated in the following manner:

- Prepare homogenous beads, cubes or cones of the support material (they should be as homogenous in size and density as possible).

- Prepare several test solutions. For example: the solutions used for application and some other critical solutions (e.g. for alginate or carrageenan demineralized water without their specific counterions).

- Put a fixed amount of support in a fixed volume of the test solution (e.g. 100 beads in 30 ml solution) and shake at a fixed speed (e.g. 100 rpm).

- Periodically (daily) measure the bead diameter, if possible the density of the support (see chapter Chapter 4) and the force to fracture the beads (see Chapter 5).

- Produce solubility graphs (Figures 1 and 2).

Biodegradability In some applications an abundant population of organisms other than the immobilized ones can be present in the system. If any of these organisms can degrade the support material the immobilized-cell process cannot be applied. To get an idea of the possibility of the biodegradability of a sup-

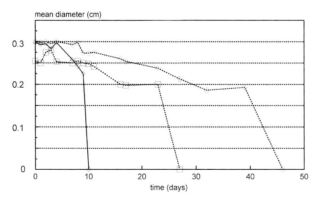

Fig. 1. Solubility of three support materials in domestic wastewater. Mean diameter as a function of time

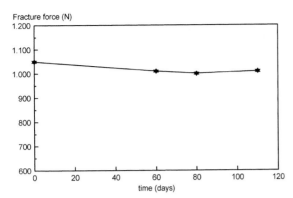

Fig. 2. Solubility of a support material. The force to fracture as a function of time

port the same procedure as described in the solubility section can be used. An extra test should, however, be done too.

Next to the solubility test, the same tests should be carried out with the addition of a toxic substance (e.g. NaN_3) to kill the cells present in the system. This serves to discriminate between solubility and biodegradability.

Furthermore, the same experiments as described above should be done in the presence of the cell population present in the application.

In Chapter 5 the measurement of mechanical stability is described in several ways. Martins dos Santos et al. (1997) described the mechanical stability of support material in bioreactors as being of crucial importance for applications of immobilized-cell technology. In this study it was found that abrasion of support material is likely to be related to fatigue of the support material. Fatigue can best be measured by **oscillation experiments** and the measurement of **force to fracture** support material during an experimental process.

Mechanical stability

Abrasion mechanism

In agitated reactors shear stresses are always present due to hydrodynamic interactions. Motion of bursting of gas bubbles, and collisions against other particles and reactor parts and the possible stresses and interactions will be discussed shortly. More information can be found in the article of Martins dos Santos.

The average energy dissipation can be calculated according to Chisti (1989):

$$\varepsilon = (F_g/mV)\ RT\ \ln(P_s/P_t)$$

Liquid shear   Liquid shear can be interpreted using the Kolmogoroff theory of isotropic turbulence. According to this theory, the energy transmitted to a reactor is first transferred to large-scale eddies which are unstable and break up into smaller eddies, and these into smaller ones and so on. Eddies with a size of the same order as the bead can not engulf the particles and will thus exert a local stress on the bead surface. Expressions for estimation of the average pressure differences and maximum pressure (stress) acting on a bead are described by Thomas (1990), Kawase and Moo-Young (1990), Cherry and Papoutsakis (1986) and Baldyga and Bourne (1995). The latter estimate the average and maximum pressure differences acting on a bead with the following expressions:

$$p(d) \sim C_p\theta(\varepsilon d_b)^{2/3}$$

$$p_{max}(d) = p(d)(d_b/L)^{-0.587}$$

Bubble shear   For air-driven reactors, the damaging interactions of gel beads with air bubbles can take place. To roughly estimate the shear stresses a particle encounters in several zones the method described by Van 't Riet and Tramper (1991) can be used.

Wall shear   On the reactor surfaces the liquid velocity is zero non-slip conditions) and thus a velocity gradient between the bulk and the surface is established. If these gradients are high enough, they may damage the support material. The average magnitude of these stresses can be estimated by (Tramper and Vlak, 1988):

$$\tau = 0.5\ \theta\ v_{ld}^2\ k_f$$

Collisions   Collisions between particles. Kusters (1991) proposed expressions to estimate the collision frequency of small particles for three different eddy sizes. At collision a compression force is induced at the point of contact. The surface is flattened and the compression force gradually decreases due to a larger contact area. The maximum is thus for the first point of contact (van den Bijgaardt, 1988). Combining the estimation of the collision frequency with the expression for an elastic head-on collision the maximum compression stress for all collisions can be calculated by (van den Bijgaardt 1988):

$$p_{max} = \theta^{2/3} \, v_{4/3} \, E_{1/3}$$

Fatigue or weariness is a process that involves the development and propagation of microcracks during fracture. Fracture starts if somewhere in the material the local stress is higher than the adhesion or cohesion forces between the bonds of its structural elements. Thus, by fracture bonds are broken and new surfaces are formed, resulting in crack growth. Near these defects or inhomogeneities the local stresses are higher than the overall stress in the material. The maximum stress is thus the stress near an irregularity which the material can sustain. As a result, fracture always starts at such an irregularity.

Fatigue of a support material

At fracture, the material around the crack relaxes and the energy can be released. Whether a crack propagates or not depends on the balance between the differential energy released with ongoing crack growth and the differential energy required to form new surfaces (Atkins & May, 1985; Luijten, 1988; Van Vliet et al., 1991).

Slow local growth of microcracks may start at much lower stresses than the experimentally observed macroscopic fracture stress. If the local stress is relieved soon it only results in growth of microcracks. If this is repeated many times many microcracks may develop, leading to a weakening of the gel structure. The bigger cracks ultimately result in abrasion of pieces of support material, while smaller microcracks result in a less translucid appearance.

The sensitivity of a material for the formation of microcracks at stresses below the macroscopic fracture stress can **not** be deduced from the normally determined stress-strain curves in uniaxial compression until fracture (see Chapter Chapter 5). Thereto the material has to be loaded in an oscillatory way as described in Chapter Chapter 5 by the oscillation experiments.

Measurement of fatigue/abrasion

Another (less efficient way) is to determine the macroscopic fracture stress of a support material during an abrasion experiment. Therefore, at time intervals pieces of the support material are subjected to the certain stress/force (Chapter Chapter 5). If a support material suffers abrasion the macroscopic fracture strain will decrease during the experiment (Figure 3).

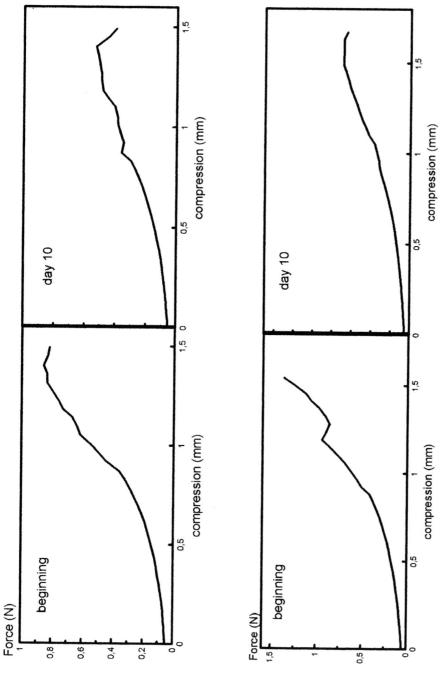

Fig. 3. Stress-strain curves of a support material at day 0 and day 10 of an abrasion experiment

Other criteria for selection of support material

- Growth of immobilized cells. It has to be taken into account that the immobilized cells also propagate microcracks in a support material, leading to a weakening of the support material.

- Attachment of heterotrophic organisms. Attached cells can hamper the diffusion of substrates to the immobilized cells or can even compete for the same substrates. This may cause a decrease in the efficiency of the overall immobilized-cell process.

- Costs. Generally the costs of synthetic supports are higher than the costs of natural support material. The stability, however, of the latter material are generally lower resulting in replacement of the support material during an experiment. A good investigation on the stability of a material is therefore crucial to calculate the actual investment and running costs of an immobilized-cell process (Leenen, 1997).

References

Atkins, A.G., May, J-M. 1985. Elastic and plastic fracture, Harwood Publs. Boston, USA

Baldyga, J., Bourne, J.R. 1995. Interpretation of turbulent mixing using fractals and multifractals. Chem. Eng. Sci. 50: 381-400.

Cherry, R.S., Papoutsakis, E.T. 1986. Hydrodynamic effects on cells in agitated tissue culture reactors. Bioprocess Engineering. 1: 29-41.

Chisti, M.Y. 1989. Airlift bioreactors. Elsevier Applied Science, New York, USA.

Kawase, Y., Moo-Young, M. 1990. Mathematical models for the design of bioreactors: application of Kolmogoroff's theory of isotropic turbulence. Chemical Engineering Journal 13: B19 B41

Kusters, K.A, 1991. The influence of turbulence on aggregation of small particles in agitated vessels. Ph.D. thesis, Technical University of Eindhoven, The Netherlands.

Leenen E.J.T.M., 1997. Nitrification by artificially immobilized cells. Model and practical system. Ph.D thesis Wageningen Agricultural University, Wageningen, The Netherlands.

Leenen, E.J.T.M., Martins dos Santos, V.A.P., Grolle, K.C.F., Tramper, J., Wijffels, R.H. 1996. Characteristics of and selection criteria for cell immobilization in wastewater treatment.Water Res. 30: 2895-2996.

Luyten, H. 1988. The rheological and fracture behaviour of Gouda cheese. Ph.D. thesis Wageningen Agricultural University, Wageningen, The Netherlands.

Martins dos Santos V.A.P, Leenen E.J.T.M., Ripoll M.M., Van der Sluis C., Van Vliet T., Tramper J., Wijffels R.H. 1997. Relevance of rheological properties of gel beads for their mechanical stability in bioreactors. Biotechnol. Bioeng. 56: 517-529.

Thomas, C.R., 1990. Problems of shear in biotechnology. In: M.A. Winkler (ed.) Chemical Engineering Problems is Biotechnology, Society Chemical Industry, Elsevier Applied Science, Doordrecht, The Netherlands, pp. 23-94.

Tramper, J., Vlak ,J.M. 1988. Bioreactor design for growth of shear-sensitive mammalian and insect cells, Upstream Processes: Equipment and Techniques : 199-228.

Riet, van 't K., Tramper, J. 1991. Basic bioreactor design. Marcel Dekker, New York, USA.

Van den Bijgaart H.J.C.M. 1988. Syneresis of rennet-induced milk gels as influenced by cheese-making parameters. Ph.D. thesis, Wageningen Agricultural University, Wageningen, The Netherlands.

Van Vliet T., Luyten, H., Walstra, P. 1991. Fracture and yielding of gels. In: E. Dickinson (ed.), Food polymers, gels and colloids. Royal Society of Chemists, Cambridge, UK: 392-403.

Abbreviations

C_p	constant according to Baldyga (-)
d_b	diameter gel beads (m)
E	Youngs modulus (N/m^s)
F_g	volumetric gas flow (m$^3 \cdot$ s^{-1})
k_f	friction factor at the wall (-)
L	size of the largest, energy-rich eddies (m) (for a bubble column these are mostly of the size of the column diameter, see Baldyga & Bourne)
m	mass of reactor contents (kg)
$p(d)$	average pressure difference (N/m^2)
$p_{max}(d)$	maximum pressure difference (N/m^2)
P_s, P_t	pressure at the sparger and the top of the columns, respectively
R	gas constant (8.314 J.mol$^{-1} \cdot$ K^{-1})
T	Temperature (K)
v_{ld}	mean liquid velocity on the downstream side of a resistance (m$^2 \cdot$ s^{-1})
Δv	velocity difference between two colliding particles (m$^2 \cdot$ s^{-1})
V	molar volume of the gas (m$^3 \cdot$ mol^{-1}) for a given temperature
ε	mean energy dissipation rate per unit mass of liquid (m^2/s^3)
ρ	medium density (Kg $\cdot$ m^{-3})
$\Delta \rho$	density difference between particle and medium (Kg $\cdot$ m^{-3})
τ	shear stress (N $\cdot$ m^{-2})

Immobilized Cells in Food Technology:
Storage Stability and Sensitivity to Contamination

CLAUDE P. CHAMPAGNE

Introduction

Vinegar was probably the first commercial food ingredient made with immobilized cell technology (ICT), and ICT is now proposed for various food processing applications. The most successful uses of ICT in the food industry are to be found in the beverages sector, particularly beer (Masschelein et al. 1994), wine production (Diviès et al. 1994) as well as for food bioconversions and the production of ingredients (Norton & Vuillemard 1994). Applications are also potentially numerous for dairy (Champagne et al. 1994b) and meat fermentations (McLoughlin & Champagne 1994).

In many processes, an important feature of ICT is the continuous or repeated use of the cells. This raises the concern over contamination. In contrast to the pharmaceutical industry, food and beverage fermentations (wine, yoghurt, beer) are often carried out on raw materials that are not sterilized, and microbial contaminants are thus bound to enter some ICT systems via the incoming ingredients. This adds to contaminations caused by air, personnel or equipment. Therefore, control of microbial contamination is an aspect that must be dealt with in industrial fermentations, and ICT processes do not escape this requirement. In the first part of this Chapter, methods of determining the susceptibility of an ICT process to contamination will be discussed. Also, some strategies to limit the development of contaminants will be examined.

While dealing with industrial applications, one must examine the reaction of the immobilized cells to process interruptions, either for plant sanitation or week-ends. The immobilized cells must be stored and kept as viable and as active as possible during these short interruptions

Claude P. Champagne, Agriculture and Agri-food Canada, Food Research and Development Center, 3600 Casavant, St. Hyacinthe, QC, J2S 8E3, Canada (*phone* +450-773-1105; *fax* + 450-773-8461; *e-mail* ChampagneC@em.agr.ca)

(a few hours to a few days). The second part of this Chapter will thus deal with methods aimed at determining the susceptibility of an ICT system to storage conditions. Additionally, ICT has been found to increase the stability of dried cultures during storage. Methods to determine if your cultures could benefit from such a treatment will be discussed.

This book's aim is to provide methods for scientists. In the area of microbiological quality control of foods, for example, there are recognized methodologies of analysis and it is possible to provide a detailed description of a given methodology. Unfortunately with respect to the determination of contamination of ICT systems or to procedures for the storage of immobilized cell, there are no such universal methodologies because of the variety of substrates, microorganisms, immobilization material and fermentation conditions that are used. Nevertheless, in order to be practical and illustrate methodological approaches, a case study will be presented, and references to other procedures will be provided.

Immobilization methods for the food industry

Methods used for immobilization of cells are described in chapters 2 to 6 and will thus not be reexamined in this Chapter. To determine which are applicable to the food and beverage industries, the reader should refer to a review done by Groboillot et al. (1994). An important aspect of ICT destined for food is the potential toxicity of the materials. As a rule, one should use immobilization matrixes that are made from compounds recognized as being safe food ingredients. Nevertheless, it is possible to use supports that have not been recognized as safe if it is clearly demonstrated that the immobilization material does not leak out or gain access to the food.

Case study: lactic cultures entrapped in alginate gels

In order to be practical and provide concrete references to the methods or strategies proposed, the Chapter will be focussed on some case studies as much as possible.

The lactic acid bacteria were selected since they are the most commonly used bacteria in food fermentations. Indeed, they are involved in dairy (yoghurt, cheese), meat (dry sausage), vegetable (sauerkraut), fruit (wine malo-lactic) and cereal (sour dough) fermentations.

There is a trend to use solid supports with cell adsorption strategies for ICT processes carried out on a large scale. However, gel beads have been more widely studied for the variety of processes available. Therefore, examples will focus on polysaccharide gel beads, in particular alginate and carrageenan.

Sensitivity to contamination

On a laboratory unit under 10 L there are reports of continuous operation of ICT for weeks (Davison & Scott 1988) or even months (Cho et al. 1988). Therefore, on a small scale, the technology is quite manageable with respect to contamination, which demonstrates the feasibility and biological stability of the processes. However, at large volumes (over 1000 L), studies have shown that the process must be examined (Kronlöf & Haikara 1991; Samajima et al. 1984; Varma et al. 1984).

Recent reports suggest that some bacteriophages do establish themselves at the surface of gel beads even at high dilution rates (Lapointe et al. 1996) and that population dynamics change at the surface of gel beads in ICT systems incorporating mixed populations (Lamboley et al. 1997). Thus, the response of an ICT bioreactor to the challenge of microbial contaminants must be determined before an industrial application can be considered.

Multiple-batch or continuous

There are basically two types of ICT systems : continuous or batch. The food and beverage fermentation industries largely operate in batch procedures, but continuous fermentations often generate high yields and are economically attractive (Samejima et al 1984). Running batch fermentations allows sanitation of the equipment, which is critical in the food sector. Consequently, introduction of continuous ICT fermentations in a plant goes against tradition, and the industrial successes of ICT in this sector often rely on multiple-batch processes. However, continuous fermentations are increasingly being considered when contamination is controlled. Some of the options available to fight contamination differ in function of the approach adopted, and both aspects will be covered when useful.

Determining the sensitivity of an ICT system to microbial contamination

In determining the sensitivity of the system to contamination, the first step is to identify the organisms involved. This can be done by:

1. conducting literature reviews,

2. examining quality control analyses,

3. isolating and identifying organisms from the finished product.

Table 1. References for the determination of various food-related floras

TECHNIQUES / METHODOLOGIES	SOURCE
Typical plate and microscopic techniques focused on food analysis. Techniques for the determination of various flora (psychrotrophic, spore-formers, acidifying, yeast and molds etc.) as well as specific pathogenic species (Salmonella, Clostridium etc.) are given. Techniques typically used by regulatory agencies. Specific sources are for (1) mycetes, (2) dairy products (3) foods in general.	(1) Arora et al. (1991) (2) Marshall (1992) (3) Vanderzant & Splittstoesser (1992)
Novel microbiological methods (lectins, PCR, immunomagnetic etc.)	Kroll et al. (1993)

Then there are two options available. The first is to follow the development of contaminants, with selective enumeration techniques, by conducting small scale fermentations that simulate the large-scale conditions. In these experiments, one must give particular care to simulate the sanitation procedures as well as the raw material treatment procedure. There are many selective enumeration techniques that are specifically related to foods, and a list of books that describe in detail the various methods appears in Table 1. The reader should refer to these excellent technical documents for specific details. Typical examples of such an approach are described in the works of Davison & Scott (1988) and Sodini et al. (1997). In the latter study, the ICT system was designed for the continuous pre-fermentation and inoculation of milk, which could subsequently be used for fresh cheese manufacture. Pasteurized milks were fed in the system for a period of 7 weeks. With such an approach, variations in the contaminating populations are inevitable, and in this instance the psychrotrophic population of the pasteurized milk fed in the system varied from 10^2 to 6×10^6

CFU / mL. The psychrotrophs did not establish themselves in the system, presumably because of the high dilution rates used in the process (D values of up to 30 h^{-1}). The advantages of the approach are that:

1. Industrial-like conditions are tested, providing a picture of the typical response of the ICT system to microbial contamination

2. Many contaminants are simultaneously tested, thus providing a rapid estimation of the potential problems. A disadvantage, however, is that not all contamination possibilities may be encountered.

Table 2. Some sources of cultures and procedures for their safe-keeping

PRODUCT / METHODOLOGIES	REFERENCE
Maintenance and propagation of microbial cultures	Bishop & Doyle (1991) Demain & Solomon (1986)
Suppliers of microbial cultures	ATCC (American Type Culture Collection), 12301 Parklawn Drive, Rockville Maryland 20852 USA BCCM (Belgian Co-ordinated Collections of microorganisms) Université Catholique de Louvain, Place Croix du Sud 3, B-1348 Louvain-la-Neuve, Belgium NCIMB (National Collections of Industrial and Marine Bacteria Ltd) 23 St. Machar Drive, Aberdeen, Scotland. U.K.

The second option is to voluntarily inoculate the ICT system with selected cultures and examine their development. This enables the scientist to better control the experiment and challenge the system with typical as well as atypical contamination situations in order to clearly define the response on the bioreactor. For this aim, one must purchase cultures from reliable suppliers or isolate some strains from products. There are numerous suppliers of cultures and a few appear in Table 2. Once the cultures are received, or the isolates obtained, it is recommended to establish methods of propagating and preserving them. Handbooks have been designed for this purpose and a few are listed in Table 2. The preparation of standardized inocula is a critical step in generating reproducible results, and care must be given to this aspect. Basically, three steps must be taken:

1. Immediately prepare stock cultures for safe-keeping. Ideally they are freeze-dried.

2. Prepare working mother cultures. Freezing a series of 1 mL cell suspensions (a medium having 10 % milk solids and 10% glycerol is good for lactic cultures) and keeping at - 40°C (or lower) is practical.

3. Determine the media and growth conditions that make possible the production of inocula with constant populations and physiological states (late logarithmic is generally preferable).

There are many examples of experiments where ICT systems were voluntarily contaminated, and methodologies can be obtained from some examples found in Table 3. In conducting such experiments, it is important to determine the appropriate contamination level. A safe approach is to use the high end of the potential contamination range.

Table 3. Studies in which ICT systems were voluntarily contaminated by microorganisms

ICT SYSTEM	CONTAMINANT	REFERENCE
Continuous inoculation of milk with lactococci	Bacteriophages	Passos et al. (1994)
Batch fermentations of milk by lactococci	Yeast	Champagne et al. (1989b)
Batch fermentations of whey by propionibacteria	Escherichia coli Staphylococcus aureus	Champagne et al. (1989a)
Continuous beer lagering	Lactobacillus brevis Pediococcus damnosus Enterobacter agglomerans Saccharomyces cerevisiae	Kronlöf & Haikara (1991)

Approaches to control contamination

In exploring ICT for a given fermentation process it is wise to develop strategies that take into account the control of unwanted microbial floras. Various approaches have been reported and some are presented in Table 4.
In examining these approaches it is possible to find three strategies.

1. Can the system be operated in a medium having naturally high inhibitory compounds? ICT processes involving the production of alcohols or acids are less prone to contamination development. Addition of inhibitors to the medium is also a possibility. This is probably the most effective method of controlling contamination. The production of vinegar on wood chippings is probably the oldest ICT, and the combina-

Table 4. Methods of preventing the establishment of contamination in an ICT bioreactor

METHOD	CONTAMINANT	PRODUCT	REFERENCE
Sulfite addition in substrate	Bacteria	Wine	Mori (1993)
Use of killer yeast	Wild yeasts	Wine	Hirotsune et al. (1987)
Low processing temperature	Non psychrotrophs	Low alcohol beer	Iersel et al., 1995
High processing temperature	Non-thermophilic organisms	High fructose syrups	Mori (1993)
High dilution rate in continuous culture	1) Psychrotrophic bacteria 2) Yeast and bacteria	1) Milk 2) Beer	1) Sodini et al. (1998) 2) Kronlöf & Haikara (1991)
Decantation of medium in multiple-batch sequences	Bacteria or yeast	Lactic or propionic acid fermentations	Champagne et al. (1989a & 1989b)
Sterilizable support	All microorganisms	Beer	
High salt	Non-halophilic organisms	Soy sauce	Nunokawa (1996)
Pre-incubation tank	Wild yeasts and bacteria	Sake	Nunokawa & Hirotsume (1993)
Antibiotics	Bacteria	Ethanol production	Nakanishi et al. (1993)
Fungicides / oil in beads	Wild yeast	Ethanol	Tanaka et al. (1994)
Low pH	Bacteria	Beer, Wine	Mori (1993)

tion of toxic substrate and product in preventing contamination is very effective.

2. Can the fermentation medium be impoverished or can extremes in pH or temperature be used? Once the appropriate biomass level is obtained, the fermentation conditions can be modified to basically generate a bio-conversion mode. An example of this approach is seen in the medium-composition work of Norton et al. (1994), who used various yeast extract supplementation levels of whey permeate for lactic acid production. As time progressed, a gradual decrease in fermentative activity was registered, but the fermentative activity could be recovered by

periodic "spikings" of the medium with appropriate yeast extract levels. The same strategy can be applied to pH and temperature, where fermentation conditions can be shifted to settings that are not necessarily ideal for the immobilized culture but that inhibit microbial contaminants. In this strategy, a cell growth or recuperation phase can be incorporated periodically to remove the stress on the cells and enable renewed growth or repairs of the damaged cells.

3. Can a contaminant-removal action be taken? This can be accomplished by high dilution rates (continuous cultures), rinsing steps (multiple-batch processes) or a physical (high temperature) or chemical (ethanol) treatment. In the two latter cases, surface contamination of gel beads can be reduced, but death of the desirable culture also occurs. A recovery period needs to be carried out after such a treatment.

The type of contaminant encountered is very important. For example, in continuous pre-fermentation of milk with lactococci on gel beads, the presence of low levels of yeast or psychrotrophic bacteria is problematic technology-wise, but generally does not pose a health threat. In this instance, the manufacturer must determine the acceptable contamination levels, obviously taking into account the legislative constraints of the products (coliform, yeast, staphylococci etc.). The same cannot be said of pathogenic species (for example *Salmonella* or *Listeria*) where a "zerotolerance" is generally practised. If the ICT system is the source of such pathogenic microbes in finished products, the bioreactor will have to be completely disinfected. This may mean killing or removing all the cells in the system and adding fresh beads (for gel-entrapped cultures) or re-seeding the support (for technologies based on surface adsorption).

Storage stability of liquid microbial cultures

Immobilization influences the immediate environment of cells, and their properties may change (McLoughlin 1994). An example worth noting is the improved stability of plasmids in immobilized cells (Yang & Shu 1996). This is obviously an important industrial aspect since it relates to the constancy of the fermentation during extended or repeated use of the cells. Another aspect of stability which is of concern to industrial users is the reaction to storage of the immobilized cultures. There are some reports of increased stability of immobilized cells in comparison to their free-cell counterparts (Asano et al. 1992; Lindsey & Yeoman 1984). An immobilized yeast culture has even maintained a good ethanol-producing activity after one year of storage at 4°C in water (Pospíchalová et al. 1990). Few immobilized cells react so well to storage and conditions must be met to maintain a high viability of the cultures. The aim of this section is to provide examples of such conditions.

Why ICT cultures need to be stored

In traditional free-cell fermentations, the plant can be stopped at will between fermentations, and the inocula are prepared according to the schedule requirements. With ICT the same cells are re-utilized. The use of ICT in continuous fermentations has the advantage of not requiring frequent process interruptions. Multiple-batch ICT operations, on the other hand, provide the possibility of stopping plant operations periodically (typically for daily sanitation or week-ends), but this poses the problem of storing the highly concentrated biomass. It is imperative to prevent death of the cells during storage, and strategies will vary as a function of the length of time that the immobilized cells need to be maintained.

Storage for a few hours or days

Microbial cells can generally be stored at 4°C for a few days without significant detrimental effects of their viability if the conditions do not generate some sort of stress. In yeast and lactic cultures, the presence of inhibitory compounds such as ethanol or organic acids increases their mortality rate. Medium pH is critical in the stability of traditional lactic starters upon storage (Ross 1980); an excess acidification of the starter of only 0.3 pH unit was shown to significantly affect the subsequent acidifying activity of the cultures. That is why it is recommended to store lactic starters within a 6.0 - 7.0 pH range. Even in these neutral conditions, free-cell lactic starters can be stored at 4°C, but activity decreases after a few days (Champagne et al. 1996). Immobilized cells seem to behave similarly to free cells in this respect. With immobilized lactic cultures, storage is possible if temperature and pH are controlled, as well as if a suitable suspension medium is included.

Temperature is probably the most important parameter. With mesophilic starters, temperatures between 1 and 7°C are required for storage, while values under 12°C are recommended for thermophilic cultures (McCoy & Leach 1997). Typically, immobilized cultures are stored at 4°C.

As a rule, keeping cell-charged gel beads in water is not recommended. Boyaval et al. (1985) showed that whey or milk were suitable storage media for alginate-entrapped cells of *Lactobacillus helveticus*. Although no data is available with respect to mesophilic cultures in the same conditions, a concern must be raised for a potential acidification of milk-based media even at 4°C. Very high cell densities are attained in gel beads and mesophilic starters will show some activity at 4°C, which did not seem to be the case with the thermophilic *L. helveticus* strain. This could result in a detrimental acidification of the medium and even milk coagulation. It would seem wise to avoid the presence of fermentable carbohydrates in the storage medium for mesophilic cultures. For short storage of carrageenan-entrapped yoghurt cultures, five media were compared (Audet et al. 1991):

1. 0.1 % peptonized water in 0.05 M KCl, pH 7.0

2. 0.1 M sodium phosphate buffer, pH 7.0

3. saline water (0.85 % NaCl), pH 7.0

4. 10 % glycerol in 0.05 M KCl

5. 10 % sorbitol in 0.05 M KCl

These media have the advantage of not having fermentable carbohydrates, but were not superior to whey or milk (Boyaval et al. 1985). The most effective storage solutions for preserving cell viability at 4°C were NaCl, glycerol and sorbitol solutions for *Streptococcus thermophilus*, and phosphate buffer and sorbitol solutions for *Lactobacillus delbrueckii* ssp. *bulgaricus* (Audet et al. 1991). This indicates the diversity of reactions of lactic cultures to environmental conditions and explains the lack of a standard methodology. When considering the storage of immobilized lactic cultures, it is thus wise to test various media and determine the best for the particular strain. This being said, in most cases some mortality or reduction in fermentation activity is to be expected after storage periods of a few days (Champagne et al., 1988). In this case, a regeneration period will need to be introduced when the cells are reutilized. As an example, Lamboley et al. (1997) stopped their continuous ICT milk inoculation/ prefermentation process for week-ends, stored the carrageenan-locust bean gum beads in peptone water (see medium 1 above) at 4°C for 64 h, and began the next week's fermentation by incubating the lactococci-containing beads for 4 hours in MRS broth. This is one approach, but regeneration could simply be carried out by allowing for a lower activity upon reutilization: dilution rates could temporarily be reduced (a few hours only) or the fermentation time of the first day could be increased in multiple-batch processes.

Finally, with respect to storage parameters, one consensus is probably pH. The pH of the storage medium must be kept in a zone considered favourable for growth. For the immobilized lactic cultures, this zone is between 6.0 and 7.0, as for the free cells.

Storage for months

Storage of ICT cultures in liquid media for extended periods requires further adjustments of the storage medium. It is necessary to prevent the cells from being active metabolically even if they are in a liquid environment. One way is to reduce the water activity (a_w) level of the solutions by adding glycerol, sugars or NaCl to the conservation solutions. Immobilized lactococci that are kept in such solutions at a_w values over 0.95 show high viability losses during their subsequent storage at 4°C (Champagne et al. 1994a). Immobilized lactococci lost only 22 % viability during a 30 day storage period at 4°C in a glycerol solution adjusted to 0.93 of a_w (approximately 30 % glycerol), but only if peptones were added to the medium as a buffering ingredient. This shows that extended storage of immobilized cells at 4°C is possible if steps are taken to adjust the pH and the water activity of the solutions. This being said, some mortality is bound to occur, and a regeneration step will be needed prior to the reutilization of the gel-entrapped cultures, as was described above.

Storage stability of freeze-dried cultures

There are three instances where immobilization of lactic cultures prior to freeze-drying was beneficial:

1. Increased survival to freeze-drying (Kearney et al. 1990a)

2. Increased survival during storage (Champagne et al. 1992)

3. Increased acidifying activity in dry sausage manufacture (Kearney et al. 1990b)

Immobilization in itself does not confer protection against mortality during freeze-drying (Champagne et al. 1996a). The treatment must be combined with various protective ingredients in the suspension medium. A variety of such ingredients are available, and the work of the Argentinian CERELA group (De Valdez et al. 1983 & 1985) provides many examples. It must be emphasized that the lactic acid bacteria show a frustrating diversity of responses to freeze-drying as a function of strain and species. This was observed not only with survival after freeze-drying, but also with stability during storage (Champagne et al. 1996b).

An advantage of alginate is that the liquid state can be recovered. Thus, the ICT approach can be applied for the benefit of stability, and the culture liquified when rehydrated.

In summary, ICT has shown many advantages with respect to stability of microbial cultures. It is nevertheless recommended not to automatically assume that ICT will increase stability of microbial cultures to storage or drying. It is hoped that the references to media and process parameters that were given in this Chapter will guide the reader in selecting the approach that would have the most potential for success for his/her particular application.

References

Arora DK, Mukerji KG & Marth EH (1991) Handbook of Applied Mycology. Foods and feeds. Marcel Dekker Inc., New-York. 621 p.

Asano H, Myoga H, Asano M & Toyao M (1992) A study of nitrification utilizing whole microorganisms immobilized by the PVA-freezing method. Wat. Sci. Tech., 26: 1037-1046.

Audet P, Paquin C & Lacroix C (1991) Effect of medium and temperature of storage on viability of lactic acid bacteria immobilized in K-carrageenan-locust bean gum gel beads. Biotechnol. Techniques. 5: 307-312.

Boyaval P, Lebrun A & Goulet J (1985). Etude de L'immobilisation de *Lactobacillus helveticus* dans des billes d'alginate de calcium Le lait. 65(649-650): 185-199.

Champagne CP, Baillargeon-Côté C & Goulet J (1988) Fermentation du lactosérum par cellules immobilisées de *Lactobacillus helveticus*. Can. Inst. Food Sci. Technol. J. 21: 403-407.

Champagne CP, Côté CB & Goulet J. (1989a) Whey fermentation by immobilized cells of *Propionibacterium shermanii*. J. Appl. Bacteriol. 66(3): 175-184.

Champagne CP, Girard F & Gardner N. (1989b) Growth of yeast contaminants in an immobilized lactic acid bacteria system. Letters Appl. Microbiol., 8(3): 207-210.

Champagne CP, Morin N, Couture R, Gagnon C, Jelen P & Lacroix C (1992) The potential of immobilized cell technology to produce freeze-dried, phage-protected cultures of *Lactococcus lactis*. Food Res. Internat., 25: 419-427

Champagne CP, Gardner N & Dugal F (1994a) Increasing the stability of immobilized *Lactococcus lactis* cultures stored at 4°C. J. Indust. Microbiol., 13: 367-371.

Champagne CP, Lacroix C & Sodini-Gallot I (1994b) Immobilized cell technologies for the dairy industry. CRC Critical Rev. Biotechnol. 14(2): 109-134.

Champagne CP, Piette M. & St. Gelais D (1995) Characteristics of lactococci cultures produced on commercial media. J. Industrial Microbiol. 15: 472-479.

Champagne CP, Mondou F, Raymond Y & Brochu E (1996a) Effect of immobilization in alginate on the stability of freeze-dried *Bifidobacterium longum*. Bioscience Microflora, 15: 9-15.

Champagne CP, Mondou F, Raymond Y & Roy D (1996b) Effect of polymers and storage temperature on the stability of freeze-dried lactic acid bacteria. Food Research Internat. 29: 555-562.

Cho HY, Yousef AE & Yang ST (1996) Continuous production of pediocin by immobilized *Pediococcus acidilactici* PO2 in a packed bed bioeactor. Appl. Microbiol. Biotechnol., 45 : 589-594.

Davison BH & Scott CD (1988) Operability and feasibility of ethanol production by immobilized *Zymomonas mobilis* in a fluidixed bed reactor. Appl. Biochem. Biotechnol. 18: 19-34.

De Valdez GF, De Giori GS, De Ruiz Holgado AP & Oliver G (1983) Comparative study of the efficiency of some additives in protecting lactic acid bacteria against freeze-drying. Cryobiology. 20: 560-566.

De Valdez GF, De Giori GS, De Ruiz Holgado AP & Oliver G (1985) Effect of drying medium on residual moisture content and viability of freeze-dried lactic acid bacteria. Appl. Environ. Microbiol. 49: 413-415.

Demain AL & Solomon NA (1986) Manual of industrial microbiology and biotechnology. American Society for Microbiology, Washington D.C. 466 p

Diviès C, Cachon R, Cavin JF & Prévost H (1994) Immobilized cell technology in wine production. CRC Critical Rev. Biotechnol., 14(2): 135-154.

Groboillot A, Boadi DK, Poncelet D & Neufeld R (1994) Immobilization of cells for application in the food industry. CRC Critical Rev. Biotechnol., 14(2): 75-108.

Hirotsune M, Nakada F, Hamachi M, Honma T (1987) J. Brew. Soc. Jpn., 82: 582.

Iersel MFM van, Meersman E., Swinkels W, Abee T, Rombouts FM (1995) Continuous production of non-alcohol beer by immobilized yeast at low temperature J. Industr. Microbiol. 14: 495-501.

Kearney L, Upton M & McLoughlin A (1990a) Enhancing the viability of *Lactobacillus plantarum* inoculum by immobilizing the cells in calcium-alginate beads incorporating cryoprotectants. Appl. Environ. Microbiol. 56: 3112-3116.

Kearney L, Upton M & McLoughlin A (1990b) Meat fermentations with immobilized lactic acid bacteria. Appl. Microbiol. Biotechnol. 33: 648-651

Kirsop BE & Doyle A (1991) Maintenance of microorganisms and cultured cells. A manual of laboratory methods, 2nd edition. Academic Press, London. 308 p.

Kroll RG, Gilmour A & Sussman M (1993). New techniques in food and beverage microbiology. Blackwell Scientific Publications, Oxford. 301 p.

Kronlöf J & Haikara A (1991) Contamination of immobilized yeast bioreactors. J. Inst. Brew., 97: 375-380.

Lamboley L, Lacroix C, Champagne CP & Vuillemard JC (1997) Continuous mixed strain mesophilic lactic starter production in supplemented whey permeate medium using immobilized cell technology. Biotechnol. Bioeng., 56 : 502-516.

Lapointe M, Champagne CP Vuillemard JC & Lacroix C (1996) Effect of dilution rate on bacteriophage development in an immobilized cells system used for continuous inoculation of lactococci in milk. J. Dairy Sci., 79(5): 767-774.

Lindsey K & Yeoman MM (1984) The viability and biosynthetic activity of cells of *Capsicum frutescens* Mill. cv. annuum immobilized in reticulate polyurethane. J. Experim. Botany, 35: 1684-1696.

McCoy DR & Leach RL (1997) Culture propagation and handling. In : Cultures for the manufacture of dairy products. Chr Hansen Inc. Milwaukee.

McLoughlin, AI (1994) Controlled release of immobilized cells as a strategy to regulate ecological competence on inocula. Biochem. Engin. Biotechnol. 51: 1-45.

McLoughlin A & Champagne CP (1994) Immobilized cells in meat fermentation. CRC Critical Rev. Biotechnol., 14(2): 179-192.

Marshall RT (1993) Standard methods for the examination of dairy products, 16 th edition. American Public Health Association, Washington. 546 p.

Masschelein CA, Ryder DS & Simon JP (1994) Immobilized cell technology in beer production. CRC Critical Rev. Biotechnol., 14(2): 155-178.

Mori A. (1993) Vinegar production in a fluidized-bed reactor with immobilized bacteria Chap17 In: Industrial application of immobilized biocatalysts, Marcel Dekker NY

Nakanishi K, Murayama H, Nagara A, Mitsui S (1993) Beer brewing using an immobilized yeast bioreactor systemChap 16 In: Industrial application of immobilized biocatalysts, Marcel Dekker NY

Nunokawa Y. (1996) Ozeki Sake Co. Personal communication

Nunokawa Y, Hirotsune M (1993) Production of soft sake by an immobilized yeast reactor system. Chap 14 In: Industrial application of immobilized biocatalysts, Marcel Dekker NY

Norton S & Vuillemard JC (1994) Food bioconversions and metabolite production using immobilized cell technology. CRC Critical Rev. Biotechnol., 14(2): 193-224.

Norton S, Lacroix C & Vuillemard JC (1994a) Reduction of yeast extract supplementation in lactic acid fermentation of whey permeate. J. Dairy Sci., 77: 2494-2508.

Passos FL, Klaenhammer TR & Swaisgood HE (1994) Response to phage infection of immobilized lactococci during acidification and inoculation of skim milk. J. Dairy Res., 61: 537-544.

Pospíchalová V, Melzoch K, Rychtera M, Basaová G (1990) Morphology and behavior of yeasts immobilized by the entrapment method. Folia microbiol, 35: 465-556.

Ross GD (1980) Observations on the effect of inoculum pH on the growth and acid production of lactic streptococci in milk Austral. J. Dairy Technol., 35 : 147-149.

Samejima H, Nagashima M, Azuma M, Noguchi S, Inuzuka K (1984) Semicommercial production of ethanol using immobilized microbial cells. Annals N.Y. Acad. Sci., 434: 394-405.

Sodini I, Boquien CY, Corrieu G & Lacroix C (1997) Microbial dynamics of co- and separately entrapped mixed cultures of mesophilic lactic acid bacteria during the continuous prefermentation of milk. Enz. Microbiol. Technol., 20: 381-388.

Tanaka H, Ohta T, Harada S, Ogbonna JC, Yagima M (1994) Appl Microbiol. Biotechnol. 41: 544-550.

Vanderzant C & Splittstoesser F (1992) Compendium of methods for the microbiological examination of foods, 3rd edition. American Public Health Association, Washington. 1219 p.

Varma R, Baliga BA, Chatterjee SK, Ghosh BK, Dalal PM, Srivastava & Karants NG (1994) Studies on the contamination of feed molasses for continuous ethanol production in an immobilized yeast cell reactor. J. Chem. Tech. Biotechnol., 34B: 111-115.

Yang ST & Shu CH (1996) Kinetics and stability of GM-CSF production by recombinant yeast cells in a fibrous-bed bioreactor. Biotechnol. Prog. 12: 449-456.

Immobilized Cells in Bioremediation

BRONAGH M. HALL and AIDEN J. MC LOUGHLIN

Introduction

A wide range of synthetic chemicals including chlorinated compounds and poly aromatic hydrocarbons (PAHs) have been disposed of in the environment as a consequence of the massive growth in industrial processes over the last century. It is only now, in the latter part of the 20th century that the environmental implications of such disposal are being assessed and strategies are emerging to bring about bioremediation, the complete mineralization of contaminants to innocuous products. Bioremediation technology exploits and optimizes the natural role of microorganisms in the transformation and mineralization of these environmental pollutants. The range of contaminated environments may include surface and subsurface soils and surface and groundwaters. Also various types of contaminating chemicals may have been disposed at a single site resulting in a complex mix of organic solvents, metals, acids and bases.

The following represent some of the discarded contaminants currently presenting problems in the environment. Chlorophenols have been used world wide as broad spectrum biocides in the lumber industry since the 1920s. Other chlorophenols have been widely used as precursors of a range of pesticides. For example, in the 1970s and 1980s world-wide production of chlorophenols was estimated at 200,000 tons per year. Five chlorinated phenols including PCP have been listed as priority pollutants by the US EPA. (Leung et al, 1997). Coal tar is a by-product of the coal gasification process used widely between 1820 and 1950. It has been estimated that 11

✉ Bronagh M. Hall, University College Dublin, Department of Industrial Microbiology, Dublin, Ireland (*phone* +353-1-7061301; *fax* +353-1-7061183; *e-mail* Bronagh.HALL@UCD.ie)
Aiden J. Mc Loughlin, University College Dublin, Department of Industrial Microbiology, Dublin, Ireland

billion tons were produced, a large proportion of which was discarded as the waste product, coal tar. This substance contains mono-aromatics and polynuclear aromatic hydrocarbons (PAHs) which have been placed on the EPA priority list because of their toxicity, mutagenicity and tendency to bioconcentrate (Findlay et al, 1995). These compounds are just a small selection of the thousands of chemicals which have been released into the soil environment.

Over time, natural degradation activities mediated by microorganisms would detoxify some of these chemicals. The objective of bioremediation is to develop affordable technology which will accelerate this process, reducing health risks and restoring the environmental site to its natural state. A wide range of microorganisms have been isolated with the ability to degrade these compounds (Leung et al, 1997; Findlay et al, 1995). However, although these organisms have performed well, in vitro success in the field has been limited. Possible reasons for such failure include inoculum handling and distribution, suppression by predators and parasites and nutrient limitation. Heterogeneity within the macroenvironment through discontinuities factors arising from fluctuations in temperature, pH and salinity throughout the soil environment can contribute to the failure (Alexander, 1994). Such factors ultimately lead to a reduction in the ecological competence of inocula - the ability of introduced microbial inoculum to survive in the environment. Possible strategies to overcome these difficulties include improved inoculum delivery systems based on immobilized cell technology which provide protection through the creation of microenvironments and facilitate controlled release of inocula to the target environment (Mc Loughlin, 1994). Immobilization of cells in polymer gels allows conservation of inocula before use enhancing stability and shelf-life (Mugnier and Jung, 1985) and enhanced inoculum distribution in the environment (Hall et al, 1998).

A range of different delivery systems has been developed based on several approaches including polymer encapsulation, entrapment and adsorption. In this chapter the most important methodologies which have applications in bioremediation are detailed.

Subprotocol 1
Alginate Encapsulation

Entrapment of cells in calcium alginate has become one of the most widely used methods for encapsulating cells. It is a fast, simple, cost-effective technique using mild conditions and readily available materials.

Alginates constitute a family of unbranched co-polymers of 1,4 linked β-D-mannuronic (M) and α-L guluronic (G) acids of widely varying composition. The residues occur in varying proportions. The acid units are arranged in blocks of homopolymeric regions (MM and GG blocks) interspersed with alternating heteropolymeric regions (Smidsrod, 1974). The relative (G)-(M) ratio affects the mechanical strength of the calcium alginate as gel strength is related to the G content which may vary from 20% to 75% depending on the source of the alginate.

The major disadvantage of the use of calcium alginate as a matrix for bead formation is its sensitivity towards chelating agents such as phosphate, citrate and magnesium ions. These ions replace the cross-linking Ca^{++} ions in the gel structure and thus disrupt the bead. Alternatively, these compounds can be used to release the cells from the bead if so desired. This factor may affect the suitability of calcium alginate for use in applications where large concentrations of such chelators are present.

▦▦ Materials

- Sodium alginate (BDH, Poole, England)
- 0.1M $CaCl_2$
- Cell culture suspension (in nutrient media or re-suspended in buffer)
- Syringe (30 ml) fitted with a 21.5 gauge needle for droplet formation or disposable pipette tip with an opening of about 1mm diameter

  Magnetic stirrer

▦▦ Procedure

1. Assemble immobilization apparatus as shown in Figure 1. Sterilise by autoclaving at 121°C for 15 min.

2. Prepare sodium alginate solution to a final concentration of 2% and sterilise by autoclaving.

3. Add microbial cell suspension to the alginate solution and mix for 5 min using a magnetic stirrer.

4. Extrude the mixture dropwise via the 30 ml syringe into an excess of 0.1M $CaCl_2$ solution.

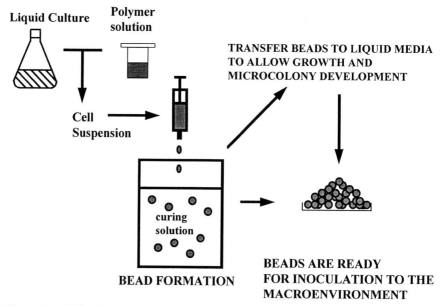

Fig. 1. Immobilization apparatus

5. Leave the beads to harden in the CaCl$_2$ solution for 30 min with gentle stirring.

6. Harvest beads and wash twice in sterile distilled water to remove excess CaCl$_2$.

▦▦ Troubleshooting

• It has been shown that autoclaving the alginate solution causes up to a 90% decrease in viscosity. However, for most practical purposes this is not a problem and is more convenient than other methods such as γ-irradiation (Leo et al, 1990).

• Preparation of the sodium alginate solution may give problems because the dry powder, being hygroscopic, will clump on the surface of the liquid. To avoid this problem dust the powder slowly to a well stirred solution in which a vortex is created. Continue to stir well for at least 30 min.

• The height between the oriface i.e. needle/pipette tip and the surface of the calcium chloride will affect the bead size and shape.

- The concentration and type of counterion used in the curing solution will influence the strength of gel formed.

- Cells can be recovered by dissolution of the alginate beads in 0.1M sodium phosphate buffer, pH 7, or 2% sodium citrate. The time needed for dissolution depends on experimental parameters such as alginate concentration, hardening time and bead cell loading.

- Vacuum filtration can be used to draw the alginate solution into the curing solution. This removes the need to manually extrude the alginate solution and may be used to partially automate bead formation.

Comments

- Increasing the concentration of the alginate solution will result in a more tightly 'cross-linked' bead. However it should be noted that the maximum workable concentration of sodium alginate is 4-5% (w/v) as above this concentration the alginate solution is highly viscous and thus difficult to extrude.

- Some buffers are not compatible with calcium alginate gels as they complex the calcium from the bead structure resulting in weakening of the bead matrix.

- The time taken for bead dissolution is a useful indicator of the level of G residues and hence the characteristics of an alginate supply.

- Beads may be transferred to fresh media and incubated to allow microbial cell growth within the bead.

- Alginate beads may be lyophilized or air-dried to enhance the stability and shelflife of the inoculum.

Subprotocol 2
Carrageenan Encapsulation

Immobilization of cells by entrapment in carrageenan gel is an alternative to alginate encapsulation. It has applications in bioremediation in situations where alginate beads may not be suitable for use (Cassidy et al, 1996).

There are three types of carrageenan, lambda (λ), kappa (κ) and iota (ι) carrageenan which are extracted from red seaweed. They consist of a backbone of alternating co-polymers of 1,3 linked β-D-galactose and 1,4-linked 3,6-anhydro-α-D-galactose. The designation, λ, κ or ι refers to the relative number and position of sulphate ester substituents on the sugars and the extent to which the 1,4 linked residues exist as the 3,6 anhydro derivative. κ-carrageenan produces firm thermally reversible gels with potassium ions, ι-carrageenan forms compliant, thermally reversible gels with calcium ions, and λ carrageenan is non-gelling. Therkelson (1993) has presented a detailed review of the production, structure and properties of carrageenans.

Materials

- κ-carrageenan type X-0909 (Genu, Denmark)
- KCl, 2% solution
- NaCl
- Cell suspension in nutrient media

Procedure

1. Prepare a -carrageenan (4%w/v) in 0.9% (w/v) NaCl. Prepare this by heating at 60°C to dissolve the polysaccharide and then maintain the solution at 40°C.

2. Assemble immobilization apparatus (as for alginate encapsulation, see previous section) containing KCl (2% w/v) at room temperature (18-20°C).

3. Mix the warm carrageenan solution with the cell suspension at a ratio of 9:1.

4. While the carrageenan is still at 40°C extrude the mixture dropwise, via a syringe, to the KCl solution.

5. Leave the beads to harden in the KCl solution for 20 min.

6. Harvest beads and wash in sterile water.

▨▨ Comments

- Carrageenan offers the greatest diversity with respect to molecular structure and range of properties of all the polysaccharides of algal origin (Guisley, 1989).

- Gelling and melting temperature of -carrageenan increases proportionally with the amount of bound potassium ions in the carrageenan. Excess potassium ions in -carrageenan can be removed in order to lower its gelling temperature (Walsh et al. 1996).

- Carrageenan can be dissolved in physiological saline at 40°C using hand-based homogenization techniques, this avoids the use of chemicals that may otherwise interfere with immobilized cells upon release.

- Gel structure is heterogeneous due probably to the thermo-gelling mechanism of the gel. This heterogeneity leads to development of irregularly shaped colonies within the gel beads (Nava Saucedo et al, 1994).

- Beads may be transferred to nutrient media to allow cell growth and microcolony development.

Subprotocol 3
Co-Immobilization with Adjuncts

Co-immobilization with nutrients, cryoprotectants or solid nutrients in a polymer gel has been shown to enhance the survival and activity of the encapsulated inoculum. A wide range of adjuvants have been used for this purpose including skimmed milk (Bashan, 1986; Fages, 1990), clay (Fravel et al, 1985) and bentonite (Trevors et al, 1992).

For bioremediation applications cells may be immobilized with an adsorbent in the gel polymer matrix. The purposes of co-immobilization are to avoid inhibition of the cells by high concentrations of toxic chemicals and to provide a suitable microenvironment for biodegradation. Adsorbents used include bentonite and activated carbon.

For example, activated carbon enhances bioremediation on its own (De Jonge et al, 1991; Ehrhardt and Rehm, 1985) and when incorporated into alginate beads (Lin and Wang, 1991). Activated carbon has a large surface area for adsorption and is thought to enhance bioremediation by adsorbing the pollutant, allowing it to be metabolized by the microorganisms

(Ehrhardt and Rehm, 1985). Addition of clays to formulations can increase cell survival and open the bead pore structure by acting as a bulking agent. Bentonite has been used to enhance inoculum survival (Van Elsas et al, 1992). Clays may bind some heavy metal ions, reducing their bioavailability and toxicity in the soil (Stotsky, 1986). However, clay amended beads may accumulate metals at levels which are toxic to the encapsulated inocula. For this reason clays may not be suitable for use in all bioremediation applications. The following protocol (Van Elsas et al, 1992) for co-immobilization with bentonite may be adapted for a range of adjuvants.

Materials

- Powdered bentonite clay (BDH Chemicals, Poole, England)
- Materials as outlined previously for alginate encapsulation.

Procedure

As for previous alginate immobilization protocol except:

1. Prepare an alginate solution (final concentration 2%) and sterilise by autoclaving.

2. Prepare a suspension of sterilized bentonite in aqueous media at 6% (w/v).

3. Add clay suspension to sterile alginate solution at a final concentration of 3% (w/v).

4. Continue steps 2-5 as for alginate immobilization protocol.

Comments

- The above protocol may be adapted for use with a range of adjuvants. For example monmortillonite clay (Fluka) or activated carbon (Sigma Aldrich Co. Ltd) may be added to the dry alginate powder at a final concentration of 1-5% and sterilized by autoclaving.

- Skimmed milk is commonly used as an adjuvant as it has been shown to enhance inoculum survival when used on its own or in conjunction with bentonite (Van Elsas et al, 1992). Prepare a skimmed milk solution in distilled water and filter through muslin to remove any lumps. Sterilise by autoclaving for 5 minutes and add to alginate solution at a final concentration of 3% (w/v).

Subprotocol 4
Immobilization with Synthetic Polymers

The first synthetic gel used to entrap living microbial cells was polyacrylamide. Polyacrylamide gels are irreversible and are characterized by their small pore size. This results in very good cell retention within the bead and is usually unsuitable for controlled release. It is an easy technique but due to the toxicity of the acrylamide monomer is not recommended for inoculum delivery but for ex situ uses (White and Thomas 1990a, 1990b, 1991). Techniques have been developed which use pre-polymerized linear polyacrylamide partially substituted with acylhydrazine groups (Figure 2) to eliminate the toxic effect.

This technique results in beads with good retention of viability. Polyacrylamide -hydrazine (PAAH) gels are very stable and do not undergo deformation as a result of environmental effects such as extreme pH, salinity or the presence of large concentrations of ions.

4.1 Entrapment of cells in cross-linked and pre-polymerized polyacrylamide

Materials

- Acrylamide (Sigma Aldrich Co.)
- N,N'- methylene-bis-acrylamide (bisacrylamide)
- Ammonium persulphate
- Coarse sieve
- 0.2 M potassium phosphate buffer, pH 7

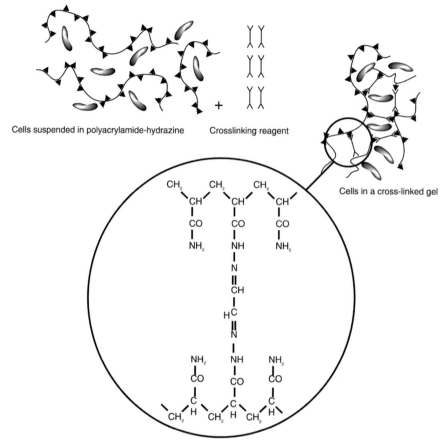

Fig. 2. Diagram of cell entrapment by polyacrylamide-hydrazine, crosslinked by di-aldehydes, Freeman and Ahronwitz (1981)

- Tetramethylethylenediamine (TEMED)
- Washed microbial cells

Procedure

1. Suspend about 5g wet weight of cells in 10 ml of distilled water and chill in ice.

2. Chill 10 ml of 0.2M potassium phosphate buffer, pH 7 in ice.

3. Add to the buffer:
   - 2.85g acrylamide
   - 0.15g bisacrylamide
   - 10 mg ammonium persulphate. Mix to dissolve these solids.

4. Immediately mix the chilled buffer solution with the chilled cell suspension, pour into glass petri dishes and cover.

5. Remove from ice and allow immobilized cells to reach room temperature. Wash gel with buffer and blot dry.

6. Gel may then be put through a coarse sieve.

7. Suspend the sieved gel in 100 ml of 0.2 M potassium phosphate buffer, pH 7.0, allow to settle and then decant the fines.

▓ ▓ Comments

- During block polymerization temperature often rises and may harm the cells. For that reason polymerization takes place on ice.

- Polymerization often appears to have been completed within a few minutes during block gel formation. From experience, it is known that at this stage a rather high content of unreacted monomer is still present. If the gel is left for maturation the amount of free monomer decreases substantially.

- The block polymer is forced through a fine net of defined mesh size, gel particle size can be controlled in this way.

- If gel size distribution is not critical the gel can be chopped into cubes with a sharp blade.

- Particles obtained by the block procedure are irregular. For use in packed-bed reactors it may be favourable to prepare spherically formed beads.

4.2 Immobilization in poly(vinylalcohol) hydrogels

Entrapment of microbial cells in synthetic hydrogels is one of the progressive approaches for the production of immobilized biomass. Cryogels are increasingly popular due to their highly porous structure which facilitates the non-hindered diffusion of solutes and dissolved gases. In particular the cryogels of poly(vinyl alcohol) (PVA) are popular due to the very high operational stability of the gel. PVA hydrogels have a highly heterogeneous and highly porous structure with macropores in the order of 0.1-1.0 μm. They have very good physicomechanical properties and their elastic and plastic properties result in little abrasive erosion. PVA gels have been successfully used in bioremediation particularly in wastewater treatment (Ariga et al, 1987); Hashimoto and Furukawa, 1986). Several methods have been developed for the encapsulation of microbial cells in PVA. One method is based on a freeze-thaw protocol (Ariga et al, 1987; Lozinsky et al, 1996) the other method involves the use of boric acid to link PVA chains (Hashimoto and Furukawa, 1986). However, the toxicity of the boric acid solution may cause problems with inoculum viability. There are variations on both these methods. For a recent review on the use of PVA for cell immobilization see Lozinsky and Plieva (1998). The method outlined here is a modification of the boric acid method. Cells are immobilized in Polyvinyl alcohol containing a small amount of calcium alginate. Contact with the boric acid solution is minimized by curing the beads in a phosphate solution therefore minimizing loss of viability. This method involves the use of boric acid to link PVA chains forming a spherical structure. This is followed by solidification of the gel bead by curing in a phosphate solution (Chen and Houng, 1997).

Materials

- Polyvinyl alcohol (Sigma Aldrich Co.)
- 0.2-1.5M monosodium phosphate solution, pH 5.0-8.0
- Saturated boric acid solution (e.g. 5.5% w/v)
- Calcium chloride
- Cell suspension concentrated and re-suspended in buffer.

Procedure

1. Mix sodium alginate (final concentration of 1% w/v) with PVA (final concentration of 15% w/v). Heat the mixture until dissolved.

2. When the PVA / alginate solution has cooled down to 30-40°C, add an equal volume of concentrated cell solution.

3. Aseptically extrude the resulting mixture dropwise into a saturated boric solution containing 1% $CaCl_2$ (immobilization apparatus, Figure 1) and gently stir for 5 min.

4. Transfer the gel beads formed to a 1.0 M sodium phosphate solution, pH 7.0 for 15 min to 2 h for hardening and to simultaneously phosphorylate the gel and disintegrate the calcium alginate polymer matrix.

5. Harvest beads by filtration and rinse the beads with sterile distilled water.

Troubleshooting

- The degree of cross-linking may be adjusted by changing the phosphate concentration. An increase in concentration will result in a lower bead porosity.

- The contact time with boric acid should be minimized to limit reduction in viability.

- The immobilization procedure should be optimized (alginate concentration of alginate, PVA, $CaCl_2$ and phosphate) depending on the physiological characteristics of the inoculum used.

Subprotocol 5
Microencapsulation

The methods detailed above are designed for the production of immobilized biomass with a diameter of 0.5-3.0 mm. Gel particles of this size are useful when one of the criteria demands protection of the microbial cells from environmental factors as outlined in the introduction. Therefore, large gel particles are most suited to in situ bioremediation and have been used successfully for this purpose. However, when ex situ bioremediation is being used especially with less toxic substances smaller gel particles may be desirable. For example, diffusional limitations can arise as a result of large bead sizes resulting in a reduction in bioremediation efficiency. Encapsulation of microbial cells can offer the advantages of immobilization with cells being retained within the gel polymer and protected from the environment while minimizing limitations of nutrients and gaseous exchange to the immobilized biomass. In this section some methods are detailed for the production of microbeads (› 0.5 μm diameter). The first method is an emulsion technique which with a little practice can be performed quickly and easily with any basic lab equipment. The low-pressure fogging nozzle technique allows the preparation of microbeads 2-100 μm diameter and can be use for large scale bead production (Stormo and Crawford, 1992; Knaebel et al, 1997).

5.1 Alginate/oil emulsion technique (beads 50-200 μm diameter)

▦ ▦ Materials

- Calcium sulphate
- Cell suspension re-suspended in appropriate buffer
- Glycerol
- Sodium alginate
- Sodium polyphosphate
- Sunflower vegetable oil (food grade)

▦ ▦ Procedure

1. Prepare an alginate solution in distilled water.

2. Dissolve 0.6g of sodium polyphosphate in 100 ml of the alginate solution and mix in a Waring blender for 15 min.

3. Weigh out 1.9g calcium sulphate, add 10 ml 50% (v/v) glycerol water mixture. Mix well and sonicate for 15 min to break down large calcium sulphate particles.

4. Add the cell suspension to the alginate solution (final concentration 2% w/v) and mix with a magnetic stirrer for 5 min.

5. Blend calcium sulphate /glycerol slurry into the alginate solution immediately before introduction into the oil phase.

6. Initiate the emulsification process by slowly transferring the alginate mix containing hydrated sodium alginate, microbial cell inoculum, sodium polyphosphate and calcium sulphate into 300 ml of Kelkin sunflower vegetable oil while vigorously mixed at a rate of 420 rpm.

7. Allow solidification of alginate beads to occur (see note 1).

8. Add 500 ml of fresh sterile distilled water to the vessel in order to break down the emulsion. Stop stirring after 5 min and refrigerate the vessel at 4°C for approximately 2 h.

9. Aseptically decant the oil and wash beads well in sterile distilled H_2O. Repeat this step with a large amount of sterile distilled water to aid removal of the oil layer (Monshipouri and Price, 1995).

Troubleshooting

- The time required for alginate bead formation may be determined by placing a small volume of the alginate mixture on a clean petri dish and observing solidification of the sample.

- The optimum speed for bead formation will depend on the type of impeller used.

- Avoid bubbling of the oil by ensuring oil is not drawn into the vortex.

- Make sure final traces of oil are removed as any remaining oil may affect microbial cell activity.

5.2 Low pressure fogging nozzle technique for microencapsulation (2-100 μm)

Materials

- Apparatus as outlined in Figure 3 including low-pressure fogging nozzle, model #052H (Sonic Environmental, Parsippany, New Jersey)

- Microbial cell suspension in re-suspended in HEPES buffer

- Sterile alginate (final concentration 3 %) in HEPES buffer/water with stirring bar.

- Sterile large-volume separation funnels

- Sterile HEPES immobilization buffer

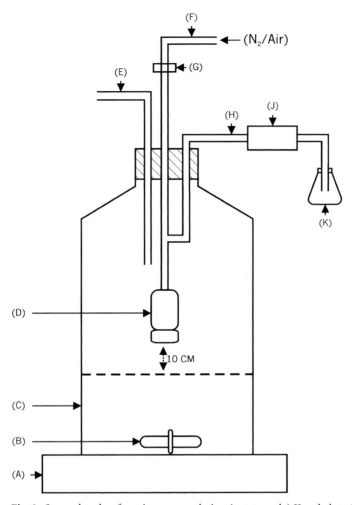

Fig. 3. Spray chamber for microencapsulation (not to scale) Knaebel et al 1997. (*A*) Stirring table (*B*) Stirring bar (*C*) Pyrex vessel containing curing solution (*D*) Low pressure fogging nozzle (*E*) Vent-line to glasswool-packed syringe (*F*) Gas line to N$_2$/air supply (*G*) In-line 0.2 μm filter (*H*) Peristaltic pump connected to fogging nozzle which delivers alginate/cell suspension at 15 ml/ min. (*J*) Peristaltic pump (*k*) Alginate/cell reservoir

Procedure

1. Add cell suspension in HEPES buffer to the alginate solution. Mix with magnetic stirrer for 5 min.

2. Place immobilization apparatus on large magnetic stirrer. Turn on mixer so solution is mixed slowly.

3. Connect tubing to gas (N_2/air) supply.

4. Connect Luer end of syringe to the vent line and direct opening into empty Erlenmeyer flask. Pack neck of flask with additional glass wool.

5. Aseptically place the free end of tubing in the alginate / cell solution. Adjust peristaltic flow rate to 15 ml min^{-1}. When the alginate / cell solution approaches the nozzle turn on gas supply to a pressure of 10 psi. A fine mist of microbeads should form and settle in the $CaCl_2$ solution which harden and sink.

6. Allow microbeads to stir slowly for 5-10 min to ensure complete solidification.

7. Aseptically decant beads through sterile separation funnel.

8. Allow beads to settle overnight at 4°C, transfer to sterile wide mouth bottles and rinse with HEPES buffer.

9. Microbeads are now ready for inoculation.

Troubleshooting

- Clumping can be avoided by optimizing the stirring speed, gas pressure and peristaltic pump flow rate.

- Bead dissolution may be carried out in phosphate buffer or tri-sodium citrate (see alginate encapsulation).

- Before placing tubing in alginate /cell solution aseptically place the tubing in sterile water and turn on peristaltic pump at rate of 15mL min^{-1}. This rinses the nozzle and tubing free of any $CaCl_2$ that may have entered the nozzle during set-up.

Comments

This protocol for microbead formation may be adapted for use with other encapsulation polymers. Hamill and Crawford (1997) detail a protocol for microencapsulation of pollutant degrading microorganisms in carrageenan and guar gum based on the above method with some modifications.

Subprotocol 6
Monitoring Microbial inoculum in the Environment

When monitoring bacteria used during in situ bioremediation it is important to be able to differentiate the strain rapidly and easily from the indigenous microbial population in order to study the fate of the introduced strain. This is critical for both risk assessment and containment. Traditional methods for tracking microbes have been based on antibiotic resistance marker systems. however there are problems associated with these systems. Firstly indigenous microorganisms may have a high level of antibiotic resistance making selection difficult. Resistant strains may be less ecologically competent than parent strains, when spontaneous mutants are used the stability of resistance may not be known following release into the environment. Bioluminescent based marker systems have many advantages – they are easy to use, visually based, highly sensitive and can be inserted in a stable form and constitutively expressed. White et al. (1996) have recently reviewed the applications of lux gene technology. Outlined here is a protocol for monitoring a lux-marked strain in soil (Flemming et al, 1994).

Materials

- A lux marked strain e.g. *P. aeruginosa* UG2Lr (Flemming et al, 1994)
- 1-decanol
- Biomedical Image Quantifier (BIQ bioview, Cambridge, UK)
- Selective media for inoculum strain e.g. TSA
- Sodium pyrophosphate

Procedure

1. Place soil sample (1g dry wt.) into 9.5 ml of sterile 0.1% (w/v) sodium pyrophosphate (adjusted to pH 7.2 with 1M HCl) shake for 1h at 200 rpm.

2. Serially dilute in 0.1% sodium pyrophosphate.

3. Plate aliquots on selective media (e.g. TSA amended with 50 μg^{-1} each of rifampicin, ampicillin and cycloheximide).

4. Invert agar plate, add a drop of 1-decanol to the lid and leave for 1 min. Bioluminescent colonies may be confirmed and enumerated using a Biomedical Image quantifier. Colonies possessing the lux AB genes produce light in the presence of 1-decanol vapour.

Conclusion

The methods detailed here are a selection of the most successful currently being used to immobilise microbial inocula and thus enhance the bioremediation of pollutant chemicals in the environment. The methodologies currently available are mostly based on the use of a single polymer gel used to immobilise a single strain of microorganism. The future success and implementation of this technology may lie in the use of more than one polymer simultaneously. This technology is already being successfully applied in other areas of inoculum delivery (Huguet et al, 1996; Huguet and Dellacherie, 1992). The co-immobilization of a microbial consortia opens the possibility of the simultaneous delivery into the environment of co-operative microbial strains which further enhance the rate of bioremediation in the environment.

References

Alexander M (1994) Inoculation. In: Alexander, M (ed) Biodegradation and bioremediation. Academic Press, pp 227-247

Ariga O, Takagi H, Nishizawa H, Sano Y (1987) Immobilization of microorganisms with PVA hardened by iterative freezing and thawing. J Ferment Technol 63:651-658

Bashan Y (1986) Alginate beads as synthetic inoculation carriers for the slow release of bacteria that affect plant growth. Appl Environ Microbiol 51: 1089-1098

Cassidy MB, Lee H, Trevors JT (1996) Environmental applications of immobilized microbial cells: A review. J Ind Microbiol 16: 79-101

Chen KC, Houng JY (1997) Cell immobilization with phosphorylated polyvinyl alcohol (PVA) gel. In: GF Bickerstaff GF (ed) Methods in biotechnology. Immobilization of Enzymes and cells, vol 1. Humana Press Inc, NJ, pp 207-216

De Jonge RJ, Breure AM, Van Andel (1991). Enhanced biodegradation of o-cresol by activated sludge in the presence of powdered activated carbon. Appl Microbiol Biotechnol 34: 683-687

Ehrhardt HM, Rehm HJ (1985) Phenol degradation by microorganisms adsorbed on activated carbon. Appl Microbiol Biotechnol 21: 32-36

Fages J (1990) An optimized process for manufacturing an *Azospirillium* Inoculant for crops. Appl Microbiol Biotechnol 32:473-478

Findlay M, Fogel S, Conway L, Taddeo A (1995) Field treatment of coal tar-contaminated soil based on results of laboratory treatability studies. In: Young LY, Cernigha CE (eds) Microbial transformation and degradation of toxic organic chemicals. Wiley-Liss Inc pp 487-513

Flemming CA, Leung KT, Lee H, Trevors JT, Greer CW (1994). Survival of lux-lac-marked biosurfactant-producing *Pseudomonas aeruginosa* UG2L in Soil monitored by nonselective plating and PCR. Appl Environ Microbiol 60: 1606-1613

Fravel DR, Marois JJ, Lumsden RD, Connick Jnr WJ (1985) Encapsulation of potential agents in an alginate-clay matrix. Phytopath 75:774-777

Freeman A, Aharonwitz Y (1981) Immobilization of microbial cells in crosslinked, pre-polymerized, linear polyacrylamide gels: antibiotic production by immobilized *Streptomyces clavuligerus* cells. Biotech Bioeng 23: 2747-2759

Guisley KB (1989) Chemical and physical properties of algal polysaccharide used for cell immobilization. Enzyme Microb Technol 11:706-716

Hall BM, Mc Loughlin AJ, Leung KT, Trevors JT, Lee H (1998) Transport and survival of alginate-encapsulated and free lux-lac marked *Pseudomonas aeruginosa* UG2Lr in soil. FEMS Microbiol Ecol 26: 51-61

Hammill TB, Crawford Rl (1997) Bacterial microencapsulation with three algal poly-saccharides. Can J Microbiol 43: 1091-1095

Hashimoto S, Furukawa K (1987) Immobilization of activated sludge by PVA-boric acid method. Biotechnol Bioeng 30: 52-59

Huguet ML, Neufeld RJ, Dellacherie E (1996) Calcium alginate beads coated with poly-cationic polymers: Comparison of chitosan and DEAE-Dextran. Process Biochem 21: 347-353

Huguet ML, Dellacherie E (1996) Calcium alginate beads coated with chitosan: effect of the structure of encapsulated materials on their release. Process Biochem 31: 745-751

Knaebel DB, Stormo KE, Crawford RL (1997) Immobilization of cells in macro- and microparticles. In: Sheehan D (ed) Methods in biotechnology, vol 2. Bioremediation Protocols. Humana Press Inc, NJ, pp 67-78

Leo WJ, Mc Loughlin AJ, Malone DM (1990) Effect of sterilisation treatments on some properties of alginate solutions and gels. Biotechnol Prog 6: 51-53

Leung KT, Errampalli D, Cassidy M, Lee H, Trevors JT, Okamura H, Bach HJ, Hall B (1997) A case study of bioremediation of polluted soil: Biodegradation and toxicity of chlorophenols in soil In: JT Trevors, Van Elsas JD, Wellington EMH (eds) Modern soil microbiology. Marcel Dekker Inc, NY, pp 577-605

Lin JE, Wang HY (1991) Degradation of pentachlorophenol by non-immobilized, im-mobilized and co-immobilized *Arthrobacter* Cells. J Ferm Bioeng 72: 311-314

Lozinsky VL, Plieva FM (1998) Poly(vinyl alcohol) cryogels employed as matrices for cell immobilization. 3. Overview of recent research and developments. Enz Microb Technol 23: 227-242

Lozinsky VL, Zubov AL, Makhlis TA (1996) Entrapment of *Zymomonas mobilis* cells into PVA-cryogel carrier in the presence of polyol cryoprotectants. In: Wijffels RH, Buitelaar RM, Bucke C, Tramper J, (Eds) Immobilized cells: basics and applications Elsevier Science B.V. Amsterdam, 112-117

Mc Loughlin AJ (1994) Controlled Release of Immobilised Cells as a Strategy to Regulate Ecological Competence of Inocula. Adv Biochem Eng/Biotech 51:1-45

Monshipouri M, Price RR (1995) Emulsification preparation of calcium alginate beads in the presence of sequesterant. J Microencap 12: 255-262

Mugnier J, Jung G (1985) Survival of bacteria and fungi in relation to water activity and the solvent properties of water in biopolymer gels. Appl Environ Microbiol 50:108-114

Nava Saucedo JE, Audras B, Jan S, Bazinet CE, Barbotin JN (1994). Factors affecting densities, distribution and growth of cells inside immobilization supports. FEMS Microbiol Rev 14: 93-98

Smidsrod (1974) Molecular basis for some physical properties of alginates in the gel state. Faraday Discuss Chem Soc 57: 263-274

Stormo KE, Crawford RL (1992) Preparation of encapsulated cells for environmental applications. Appl Environ Microbiol 58: 727-730

Stotsky G (1986) Influence of soil mineral colloids on metabolic processes, growth, adhesion and ecology of microbes and viruses. Interactions of soil minerals with natural organics and microbes. Soil Sci. Soc. Amer. Madison, WI, USA. Special publication no. 17, pp 305-428

Therkelson GH (1993) Carrageenan. In: Whistler RL, BeMiller JN (eds) Industrial gums, 3rd ed. Academic Press, NY, pp 145-180

Trevors JT, Van Elsas JD, Lee H,Van Overbeek LS (1992) Use of alginate and other carriers for encapsulation of microbial cells for use in soil. Microb Rel 1:61-69

Van Elsas JD, Trevors JT, Jain D, Wolters, Heijen CE, Van Overbeek (1992) Survival of, and root colonization by, alginate-encapsulated *Pseudomonas fluorescens* cells following introduction into soil. Biol Fert Soils 14:14-22

Walsh PK, Isdell FV, Noone SM, O Donovan MG, Malone DM (1996) Growth patterns of *saccharomyces cerevisiae* microcolonies in alginate and carrageenan gel particles: Effect of physical and chemical properties of gels. Enz Microb Tech 18: 366-372

White GF, Thomas ORT (1990b) Immobilization of the surfactant degrading bacterium *Pseudomonas* C12B in polyacrylamide gel. II. Optimizing SDS-degrading activity and stability. Enzyme Microb Technol 12:969-975

White D, Leifert C, Ryder MH, Killham K (1996). Lux gene technology-a strategy to optimize biological control of soil-borne diseases. New Phytol 133: 173-181

White FG, Thomas ORT (1990a) Immobilization of the surfactant-degrading bacterium *Pseudomonas* C12B in polyacrylamide gel beads: 1. Effect of immobilization on the primary and ultimate biodegradation of SDS, and redistribution of bacteria within beads during use. Enzyme Microb Technol 12: 697-705

White GF, Thomas ORT (1991) Immobilization of the surfactant degrading bacterium *Pseudomonas* C12B in polyacrylamide gel. III. Biodegradation specificity for raw surfactants and industrial wastes. Enzyme Microb Technol, 13: 338-343

Suppliers

Alginate by:
BDH Laboratory suppliers
Poole BH15 1TD
England
Phone: +44 (0202) 669700
Fax: +44 (1445) 558586

Carrageenan by:
Genu
Copenhagen Pectin A/S
DK-4623 Lille Skensved
Denmark
phone: +45-56165616
fax: +45-56169446

Co-immobilisation with adjuncts by:
Fluka chemie AG
P.O. box 260
CH-9471 Buchs
Switzerland
Phone: +44 (81) 7552511
Fax: +44(81) 756 5449

And:
Sigma-Aldrich Co Ltd
Europe
Phone: 1-800-200-888
Fax: 1-800-600-222
Email_elOrders@vms.sial.com

Plasmid Stability in Immobilized Cells

JEAN-NOËL BARBOTIN

Introduction

Genetic engineering provides the means to design and construct whole cells which overproduce desired and valuable proteins and metabolites. However, during the development of a recombinant strain (mostly bacteria and yeasts) to be used in bioreactor, a major concern is the plasmid instability which currently occurs in continuous processes (Imanaka and Aiba, 1981; Ensley, 1986). Construction of stable plasmids and strategies of control of the environment have been carried out to enhance plasmid stability in fermentation cultures (Kobayashi et al, 1991; Kumar et al, 1991; Wu and Wood, 1994). On the other hand, immobilized cells present a number of advantages over free suspension cells. In particular, it is possible to increase fermentor productivity by substantially increasing the immobilized population density and by using high flow rates in continuous operations. The uncoupling of growth and production phases and the separation of cells from media also facilitate downstream processing and a higher cell viability is often observed.

Immobilization can also give rise to a higher retention of plasmid-bearing cells and thus delay overgrowth by the corresponding plasmid-free cells (Kumar and Schügerl, 1990; Barbotin, 1994; Zhang et al, 1996; Barbotin et al, 1998).

Jean-Noël Barbotin, Université de Picardie Jules Verne, Laboratoire de Génie Cellulaire, UPRES-A CNRS 6022, 33 rue Saint-Leu, Amiens, 80039, France
(*phone* + 33-3-22-82-75-95; *fax* +33-3-22-82-75-95;
e-mail Jean-Noel.Barbotin@sc.u-picardie.fr)

Outline

Plasmid stability in immobilized cell cultures

The stability of a plasmid is defined as the capacity of the host cell to retain the integral structures of the plasmid, to allow its expression and to maintain at least one copy of the plasmid per cell. Microorganisms carrying plasmids are susceptible to both structural and segregational instability. Structural instability arises from physical changes in the plasmid DNA structure such as deletions, insertions and rearrangements. Segregational stability is due to improper partitioning of the plasmid between daughter cells during cell division. Furthermore, a competitive instability may also occur, due to the growth advantage of plasmid-free cells over plasmid-carrying cells. The application of antibiotic selection pressure in the reactor to suppress the growth of plasmid-free cells is a common method to overcome the drawbacks of recombinant plasmid instability, but such a method is not appropriate in industrial-scale cultivation. For free cell cultures, different strategies have been developed such as the use of active partition, postsegregational killer locus or two-stage reactors and cyclic environmental changes. Immobilization of cells which may reduce the growth rates within the matrix has also been proposed by some authors as a convenient procedure. Inloes et al, (1983) first reported increased plasmid stability in *E. coli* cultures immobilized in a non-selective hollow fiber system. Similarly, Georgiou et al, (1985) have used *E. coli* pKK entrapped in alginate gel beads in a resting state to continuously produce a target protein. This stabilizing effect has also been observed by other authors using different kind of matrices for cell immobilization such as agarose (Birnbaum et al, 1988; Ariga et al, 1991), calcium alginate (Sode et al, 1988; Vieth, 1989; Roca et al, 1996), carrageenan (De Taxis du Poët et al, 1986; Nasri et al, 1988; Ryan and Parulekar, 1991; Dincbas et al, 1993), cotton cloth (Joshi and Yamazaki, 1987; Zhang et al, 1997a), gelatine beads (Walls and Gainer, 1989), glass beads (Shu and Yang, 1996), polyacrylamide/hydrazide (Kanayama et al, 1988), polyvinyl alcohol (Ariga et al, 1997), silicone foam (Oriel, 1988). In non-selective media, factors affecting plasmid stability in immobilized *E. coli* cells may include plasmid characteristics (Sayadi et al, 1988), host strains (Nasri et al, 1987), growth rate (De Taxis du Poët et al, 1987), plasmid copy number (Sayadi et al, 1989), dilution rate (Sayadi et al, 1989), nutrient limitations (Sayadi et al, 1989), oxygen limitations (Marin-Iniesta et al, 1988; Huang et al, 1990 a,b). Other kinds of immobilized recombinant cells have been tested such as *Lactococcus lactis* (D'Angio et al, 1994), *Pediococcus acidilactici* (Huang et al, 1996), *Pseudomonas putida* (Karbasi et al, 1996), *Bacillus subtilis* (Craynest et al, 1996), *Saccharomyces cerevisiae* (Walls and Gainer, 1989, 1991; Jeong et al, 1991; Chang et al, 1996; Roca et al, 1996; Shu and Yang, 1996; Yang and Shu, 1996; Huang et al, 1997; Zhang et al, 1997a).

Modelling plasmid instability kinetics in recombinant cells can also provide valuable information and mathematical simulations may be useful tools in the prediction of the extent of plasmid stability in free and immobilized cells (Lee and Bailey, 1984; Sardonini and DiBiasio, 1987; Mosrati et al, 1993; Zhang et al, 1997b).

Plasmid stability in biofilm cultures

In biofilm cultures, it has been shown that plasmid transfer may reduce the probability of plasmid loss (Saye et al, 1987). However studies of plasmid stability in biofilm cultures have been performed to understand the fate of recombinant strain released in an open environment. In this way, Huang et al, (1993, 1994) have observed a decrease in plasmid stability in the biofilm cultures of *E.coli* DH5(pMJR1750), which was attributed to plasmid copy number differences between suspended and biofilm cultures and to a metabolic demand for the extracellular polysaccharide (EP) production. These authors have suggested that in the biofilm cultures, cells preferentially channel energy to synthetize and excrete EP rather than expressing a heterologous plasmid-encoded protein. In transconjugant *Burkholderia cepacia* (TOM$_{31c}$) biofilm cultures, Sharp et al, (1998) showed that the activity and expression of the Tom pathway measured was significantly less than that found in suspended cultures at comparable growth rates.

Materials

Supports for immobilization

Kappa-carrageenan (E407) was obtained from Satia, CECA, Carentan, France.

Media

For *E. coli* (B, BZ18, W3101, W3110) cultures, the medium used throughout the experiments was Luria-Bertani (LB): 10 g $\cdot$ L^{-1} pancreatic hydrolysate of casein (BioMerieux, France), 5 g $\cdot$ L^{-1} yeast extract (BioMerieux, France), 5 g $\cdot$ L^{-1} NaCl. The pH was adjusted to 7.3 before sterilization. Bacteria were preincubated overnight in LB medium containing 150 µg $\cdot$ mL^{-1} ampicillin.

When M9 minimal medium was used (pH 7.3), it was supplemented with 1 g · L^{-1} glucose, casaminoacids and 0.05 g · L^{-1} thiamin. For immobilized cells, LB medium was supplemented with 0.1M KCl to ensure mechanical stability of the gel bead.

Strains and plasmid vectors

In the present study, we have used recombinant *E. coli* strains bearing a plasmid (derivative from pBR322 as pTG201, pTG205 and pTG206) carrying the *Pseudomonas putida xyl*E gene encoding catechol 2,3-dioxygenase (C23O). In the case of pTG201, C23O is under the transcriptional control of λP_R and C_I 857 repressor (Sayadi et al, 1987). When using *E. coli* W3110 harboring pTG205, C23O is under the control of the trp promoter (Berry et al, 1990). All the plasmids tested were generous gifts from Transgene S.A., France.

Procedure

The following examples concern only *E. coli* cells entrapped within carrageenan gel beads.

Cell immobilization within carrageenan gel beads

A bacterial suspension from a preculture (4.2 mL) was mixed with 36 mL of 2.2% (w/v) κ-carrageenan sterile solution at 42°C. This mixture was pumped (peristaltic pump, Minipuls 3, Gilson, France) through a needle at a rate of approximately 2 mL · min^{-1} and dropped into 150 mL of a sterile 0.3 M KCl solution. Then, gel beads (about 3500) with an average diameter of 3 mm were formed. After 30 min the solution was decanted and the beads were washed with a 9g · L^{-1} NaCl solution.

Continuous immobilized cell culture

Small reactor   Culture experiments were done in 50 mL volumes in 100 mL glass vessels that were maintained at 37°C, with aeration at 170 mL · min^{-1} in absence of selection pressure. The chemostat was stirred continuously at 250 rpm. For free cultures, experiments were generally started with an overnight batch culture in 20 mL LB medium containing antibiotic (ampicillin 50µg · mL^{-1}). The chemostat was then inoculated with 0.4 mL of this culture. After batch growth for 4-5 h, fresh medium flow was started at the desired dilution rate. For immobilized culture, the gel beads formed (10 mL) were transferred to the chemostat containing 40 mL of growth medium (LB), and immediately a continuous flow was initiated at a dilution rate higher than the maximum growth rate.

A 1L fermentor (Setric Genie Industriel, France) equipped with automatic controls for dissolved oxygen concentration, temperature and pH was used. Agitation rate was controlled at 50 rpm and the dilution rates were regulated at a value of 1.5 times or 0.7 times maximum growth rate for immobilized and free bacterial cultures respectively.

Fermentor

In the first stage, immobilized recombinant cells were grown in a repressed state (31°C in the case of pTG201 or in the presence of tryptophan in the case of pTG205) to prevent loss of plasmid and the escaped cells were continuously transferred to the second stage where derepression was induced (controlling the temperature at 42°C or by the presence of 30 µg mL^{-1} 3-β-indolyl acrylic acid-added from a fresh solution of 10 mg · mL^{-1} in ethanol, respectively). The dilution rate of the second stage was adjusted by adding M9 medium or by increasing the reactor volume.

Two-stage reactor

Determination of viable cell concentration

From the gel beads prepared for cultures, at least 8 gel beads (small reactor) or 20 beads (fermentor) were weighed and then dissolved at 37°C by addition of 4 mL sodium citrate (10 g · L^{-1}).The sample was successively diluted tenfold using 64 mM phosphate buffer (pH 7.5) and 50-200 mL aliquots of the appropriate dilution were spread onto LB agar plates. The plates were incubated at 37°C until colonies were visible and the viable cell concentration was determined from the number of colonies on the plates. Colonies carrying the plasmid were quickly detected by spraying the plates with an aqueous solution of 0.5 M catechol. Colonies expressing the gene for catechol 2,3-dioxygenase became yellow due to the conversion of catechol to 2-hydroxymuconic semialdehyde, while others remained white.

Assays of catechol 2,3-dioxygenase activity

Kinetic studies of catechol conversion into 2-hydroxymuconic semialdehyde were performed with a spectrophotometer. The absorbance of the yellow product occurs at 375 nm, which is different from that of catechol which absorbs at 275 nm (Zukowski et al, 1983). The activity measurements were made in a 100mL reactor (thermostated at 30°C) from which the solution of 27 mL Hepes buffer (0.083 M) and 3 mL catechol (10^{-3} M) was continuously circulated through a flow cuvette, then 0.1mL sample was added to the solution (Dhulster et al, 1984). Enzyme activity was expressed as specific activity. One unit of enzyme activity is defined as 1×10^{-9} OD per viable bacterium carrying plasmid per minute.

Gel electrophoresis of nucleic acids

Plasmid DNA was extracted from the cells according to Birnboim and Doly (1979). Horizontal 0.8% (w/v) agarose gels were prepared with Tris buffer (Tris/acetate 40 mM pH 7.7, sodium acetate 20 mM, EDTA 1mM) and electrophoresis carried out at a constant voltage of 60 V for 4-6h. The gels were stained in ethidium bromide ($0.5\mu g \cdot mL^{-1}$) for 30 min, then excess stain was removed by washing in deionized water (45 min) and photographed under UV light (302 nm) with Polaroid film.

Plasmid copy number

Chromosomal and plasmid DNA were extracted by the method of Eckhardt (1978) modified by Projan et al, (1983). The cells were treated with lysozyme, ribonuclease, SDS and proteinase K at 37°C and electrophoresis of the lysate carried out in vertical agarose gel. DNA content of each band was determined by microdensitometer tracing of photographic negatives. The number of plasmid copies per cell was calculated by using the equation described by Projan et al, (1983): $Cp=DpMc/DcMp$. Cp indicates the plasmid copies per cell, Dp and Dc are the amounts of plasmid and chromosome DNA in the gel and Mp and Mc are the molecular masses of the plasmid and chromosome respectively.

Microscopic observations

Samples were fixed in 3% glutaraldehyde in 0.1M cacodylate buffer (supplemented with 0.1 M KCl) for 2 h at ambient temperature then washed with the same buffer and postfixed with 1% OsO_4 for 2 h at ambient temperature. For scanning electron microscopy, the specimens were then dehydrated and sputter coated in gold. For transmission electron microscopy, the specimens were dehydrated in alcohol and propylene oxide and then embedded in Epon 812 resin. Ultrathin sections were prepared with an ultramicrotome and the thin sections were double stained with uranyl acetate and lead citrate.

Results

Plasmid stability in continuous cultures

The enhanced stability of the recombinant plasmid pTG201 containing the *xylE* gene in immobilized *E. coli* within carrageenan gel beads in the absence of antibiotic selection has been demonstrated. In free cell continuous culture, loss of plasmid was detected after 25-30 generations and after prolonged incubation the proportion of plasmid-containing cells (P^+ cells) gradually decreased. In contrast, when the strain was cultivated in immobilized cell cultures, pTG201 was completely stable for 300 generations (Nasri et al, 1987).These observations have been confirmed with other *E. coli* strains (De Taxis du Poët et al, 1987; Sayadi et al, 1988) and other pBR322-related plasmids (Sayadi et al, 1989). For immobilized cells, enzyme production and plasmid copy number have been maintained constantly at high level for 100 generations (Sayadi et al, 1989). Even under glucose, nitrogen, or phosphate limitation, immobilization enhanced the stability of plasmid (Sayadi et al, 1989). Plasmid pTG201 stability increased with increasing inoculum size in the gel and larger inoculum reduced the number of cell divisions required to fill the cavities in the carrageenan gel bead (Berry et al, 1988). In addition, because of the large inoculum, only few cavities were contaminated by plasmid free cells, so there was little competition between P^+ and P^- cells. This was consistent in terms of the generation number of the commonly found lag phase, which occurs before plasmid instability begins. It can be concluded that the increased plasmid stability in immobilized cells may have resulted from the mechanical properties of the gel bead system that allows only a limited number of cell divisions to occur in the microcolonies within the matrix before cell leakage. This was supported by microscopical observations showing, when a greater inoculum density was used, a bacterial growth limited to a peripherical layer of the bead (Berry et al, 1988). Different inoculum densities (from 2×10^3 to 2×10^{10} cells $\cdot$ mL^{-1}) have been used in immobilized cell cultures but the final biomass inside the gel beads after 24 h culture was of the same order (above 2×10^{10} cells $\cdot$ mL$^{-1)}$). However, the length of the period corresponding to the decrease of the cells bearing plasmid appears to be a function of inoculum size. The plasmid was found to be more stable with a high inoculum size. It has been calculated (Berry et al, 1988) that 26 cell divisions were required to obtain a steady state in the case of a small inoculum density (4.7×10^3 cells $\cdot$ mL^{-1}). In contrast, with a high inoculum density (2×10^{10} cells $\cdot$ mL^{-1}), only 4-5 divisions were required. Furthermore, the role played by oxygen (Marin Iniesta et al, 1988) has been emphasized showing that the depth of O_2 penetration in the gel bead decreased with increasing cell growth (Huang et al, 1990a). The existence of a steep gradient in viable cell density in gel-entrapped bacteria has also been observed (Briasco et al, 1990). The effects of anaerobic conditions have also been studied with *E. coli* B (pTG201) and *E. coli* HB101 (pKBF367-11) indicating, at high cellular density in the gel beads, maintenance of plasmid and stabilization of the plasmid copy number (Ollagnon et al, 1993).

Cultures of *E. coli* in a two-stage chemostat

The phenomena of cell leakage have been exploited in the development of two-stage reactors (Sayadi et al, 1987; Berry et al, 1990). A two-stage chemostat was used to separate the phases of growth and product synthesis. Immobilization was used to produce stable released cells in the first stage in the presence of 20 mg $\cdot$ mL^{-1} tryptophan. These leaked cells could not divide in the reactor because the residence time was much shorter than the minimum doubling time. The immobilized cells can be considered as a reservoir of plasmid-bearing cells which are saved from competition with plasmid-free cells. After 150 h of culture only 18% of plasmid-free cells were detected. The escaped cells were continuously transferred to the second stage where IAA was added and where the plasmid was found to be more stable when a high dilution rate was used. For example (Berry et al, 1990), after 100 h when the dilution rate was 1.3 h^{-1} and the volume of the reactor in the second stage 50 mL, the volumetric production rate of catechol 2,3-dioxygenase was around 530 units L^{-1} h^{-1} (the volumetric production rate was increased 15-fold as compared to free cell cultures).

Comments

Immobilization should be considered as a general and simple sucessful technique to improve plasmid stability of recombinant cells in continuous cultures. The presence of a foreign plasmid in a cell might cause reductions in the growth rate and/or in some specific enzyme activities linked to transcription processes. It can be assumed that the mechanical protection of the cells by a three-dimensional matrix may contribute also to the functional stability. Cell-cell interactions and physiological changes of immobilized cells may also contribute to the stabilization process.

Acknowledgements. Thanks to Drs Berry F, Craynest M, De Taxis du Poët P, Dhulster P, Huang J, Mater D, Nasri M, Nava Saucedo JE, Sayadi S, Ollagnon G, Thomas D and Truffaut N for their contributions.

References

Ariga O, Ando Y, Fujishita Y, Watari T, Sano Y, (1991) Production of thermophilic α – amylase using immobilized transformed *Escherichia coli* by addition of glycine. J Ferment Bioeng 71:397-402

Ariga O, Toyofuku H, Minegishi I, Hattori T, Sano Y, Nagura M (1997) Efficient production of recombinant enzymes using PVA-encapsulated bacteria. J Ferment Bioeng 84:553-557

Barbotin JN (1994) Immobilization of recombinant bacteria: a strategy to improve plasmid stability. Ann NY Acad Sci 721:303-309

Barbotin JN, Mater DDG, Craynest M, Nava Saucedo JE, Truffaut N, Thomas D (1998) Immobilized cells: plasmid stability and plasmid transfer. In Ballesteros A, Plou PJ, Iborra JL, Halling P (eds) Stability and Stabilization of Biocatalysts, Elsevier, Amsterdam pp 591-602

Berry F, Sayadi S, Nasri M, Barbotin JN, Thomas D (1988) Effect of growing conditions of recombinant *E. coli* in carrageenan gel beads upon biomass production and plasmid stability. Biotechnol Lett 10:619-624

Berry F, Sayadi S, Nasri M, Thomas D, Barbotin JN (1990) Immobilized and free cell continuous cultures of a recombinant *E. coli* producing catechol 2,3-dioxygenase in a two-stage chemostat: improvement of plasmid stability. J Biotechnol 16:199-210

Birnbaum S, Bulow L, Hardy K, Mosbach K (1988) Production and release of human proinsulin by recombinant *Escherichia coli* immobilized in agarose microbeads. Enzyme Microb Technol 10:601-605

Birnboim HC, Doly J (1979) A rapid alkaline extraction procedure for screening recombinant plasmid DNA. Nucleic Acid Res 7:1513-1523

Briasco CA, Barbotin JN, Thomas D (1990) Spatial distribution of viable cell concentration and plasmid stability in gel-immobilized recombinant *E. coli* . In De Bont JAM, Visser J, Mattiasson B, Tramper J (eds) Physiology of immobilized cells, Elsevier, Amsterdam, pp 393-398

Chang HN, Seong GH, Yoo IK, Park JK, Seo JH (1996) Microencapsulation of recombinant *Saccharomyces cerevisiae* cells with invertase activity in liquid-core alginate capsules. Biotechnol Bioeng 51:157-162

Craynest M, Barbotin JN, Truffaut N, Thomas D (1996) Stability of plasmid pHV1431 in free and immobilized cultures: effect of temperature. Ann NY Acad Sci 782:311-322

D'Angio C, Beal C, Boquien CY, Corrieu G (1994) Influence of dilution rate and cell immobilization on plasmid stability during continuous cultures of recombinant strains of *Lactococcus lactis* subsp. *lactis*. J Biotechnol 32:87-95

De Taxis du Poët P,. Dhulster P, Barbotin JN, Thomas D (1986) Plasmid inheritability and biomass production: comparison between free and immobilized cell cultures of *Escherichia coli* BZ18 (pTG201) without selection pressure. J Bacteriol 165:871-877

De Taxis du Poët P, Arcand Y, Bernier Jr R, Barbotin JN, D. Thomas (1987) Plasmid stability in immobilized and free recombinant *Escherichia coli* JM105 (pKK223-200): importance of oxygen diffusion, growth rate and plasmid copy number. Appl Environ Microbiol 53:1548-1555

Dhulster P, Barbotin JN, Thomas D (1984) Culture and bioconversion use of plasmid-harboring strain of immobilized E. coli. Appl Microbiol Biotechnol 20:87-93

Dincbas V, Hortascu A, Camurdan A (1993) Plasmid stability in immobilized mixed cultures of recombinant *Escherichia coli*. Biotechnol Prog 9:218-220

Eckhardt T (1978) A rapid method for the identification of plasmid desoxyribonucleic acid in bacteria. Plasmid 1:584-588

Ensley BD (1986) Stability of recombinant plasmids in industrial microorganisms. Crit Rev Biotechnol 4:263-277

Georgiou G, Chalmers JJ, Shuler ML, Wilson DB (1985) Continuous immobilized recombinant protein production from E. coli capable of selective protein excretion: a feasibility study. Biotechnol Prog 1:75-79

Huang CT, Peretti SW, Bryers JD (1993) Plasmid retention and gene expression in suspended and biofilm cultures of recombinant *Escherichia coli* DH5α (pMJR1750). Biotechnol Bioeng 41:211-220

Huang CT, Peretti SW, Bryers JD (1994) Effect of medium carbon-to-nitrogen ratio on biofilm formation and plasmid stability. Biotechnol Bioeng 44:329-336

Huang J, Dhulster P, Thomas D, Barbotin JN (1990a) Agitation rate effects on plasmid stability in immobilized and free-cell continuous cultures of recombinant *E. coli*. Enzyme Microb Technol 12:933-939

Huang J, Hooijmans CM, Briasco CA, Geraats SGM, Luyben KCM, Thomas D, Barbotin JN (1990b) Effect of free cell growth parameters on oxygen concentration profiles in gel immobilized recombinant *Escherichia coli*. Appl Microbiol Biotechnol 33:619-623

Huang J, Lacroix C, Daba H, Simard RE (1996) Pediocin 5 production and plasmid stability during continuous free and immobilized cell cultures of *Pediococcus acidilactici* UL5. J Appl Bacteriol 80:635-644

Huang YL, Shu CH, Yang ST (1997) Kinetics and modeling of GM-CSF production by recombinant yeast in a three-phase fluidized bed bioreactor. Biotechnol Bioeng 53:470-477

Imanaka T, Aiba S (1981) A perspective on the application of genetic engineering: stability of recombinant plasmid. Ann NY Acad Sci 369:1-14

Inloes DS, Smith WJ, Taylor DP, Cohen SN, Michaelis AS, Robertson CR (1983) Hollow fiber membrane bioreactors using immobilized E. coli. for protein synthesis. Biotechnol Bioeng 25:2653-2681

Jeong YS, Vieth WR, Matsuura T (1991) Fermentation of lactose to ethanol with recombinant yeast in an immobilized yeast membrane bioreactor. Biotechnol Bioeng 37:587-590

Joshi S, Yamazaki H (1987) Stability of pBR322 in *Escherichia coli* immobilized on cotton cloth during use as resident inoculum. Biotechnol Lett 9:825-830

Kanayama H, Sode K, Karube I (1988) Continuous hydrogen evolution by immobilized recombinant *E. coli* using a biorector. Biotechnol Bioeng 32:396-399

Karbasi M, Asilonu E, Keshavarz T (1996) Improved stability of a naturally occuring TOL plasmid in *Pseudomonas putida* by immobilization. Prog Biotechnol 11:458-463

Kobayashi M, Kurusu Y, Yukawa H (1991) High-expression of a target gene and high-stability of the plasmid. Appl Biochem Biotechnol 27:145-161

Kumar PK, Schügerl K (1990) Immobilization of genetically engineered cells: a new strategy for higher plasmid stability. J Biotechnol 14:255-272

Kumar PK, Maschke HE, Friechs K, Schügerl K (1991) Strategies for improving plasmid stability in genetically modified bacteria in bioreactors. Trends Biotechnol 9:279-284

Lee SB, Bailey JE (1984) Analysis of growth rate effects on productivity of recombinant *E. coli* populations using molecular mechanism models. Biotechnol Bioeng 26:66-73

Marin-Iniesta F, Nasri M, Dhulster P, Barbotin JN, Thomas D (1988) Influence of oxygen supply on the stability of recombinant plasmid pTG201 in immobilized *E. coli* cells. Appl Microbiol Biotechnol 28:455-462

Mosrati R, Nancib N, Boudrant J (1993) Variation and modelling of the probability of plasmid loss as a function of growth rate of plasmid-bearing cells of *E. coli* during continuous culture. Biotechnol Bioeng 41:395-404

Nasri M, Sayadi S, Barbotin JN, Dhulster P, Thomas D (1987) Influence of immobilization on the stability of pTG201 recombinant plasmid in some strains of *E. coli*. Appl Environ Microbiol 53:740-744

Nasri M, Berry F, Sayadi S, Thomas D, Barbotin JN (1988) Stability fluctuations of plasmid-bearing cells: immobilization effects. J Gen Microbiol 134:2325-2331

Ollagnon G, Truffaut N, Thomas D, Barbotin JN (1993) Effects of anaerobic conditions on biomass production and plasmid stability in an immobilized recombinant *Escherichia coli*. Biofouling 6:317-331

Orlei P (1900) Immobilization of recombinant *Escherichia coli* in silicone polymer beads. Enzyme Microb Technol 10:518-523

Projan SJ, Carleton S, Novick RP (1983) Determination of plasmid copy number by fluorescence densitometry. Plasmid 9:182-190

Roca E, Meinander N, Hahn-Hägerdal B (1996) Xylitol production by immobilized recombinant *Saccharomyces cerevisiae* in a continuous packed-bed bioreactor. Biotechnol Bioeng 51:317-326

Ryan W, Parulekar SJ (1991) Immobilization of *Escherichia coli* JM103(pUC8) in κ-carrageenan coupled with recombinant protein release by in situ cell membrane permeabilization. Biotechnol Prog 34:99-110

Sardonini CA, DiBiasio D (1987) A model for growth of *Saccharomyces cerevisiae* containing a recombinant plasmid in selective media. Biotechnol Bioeng 27:1668-1674

Sayadi S, Nasri M, Berry F, Barbotin JN, Thomas D (1987) Effect of temperature on the stability of plasmid pTG201 and productivity of *xyl*E gene product in *Escherichia coli*: development of a two-stage chemostat with free and immobilized cells. J Gen Microbiol 133:1901-1908

Sayadi S, Berry F, Nasri M, Barbotin JN, Thomas D (1988) Increased stability of pBR322-related plasmids in *Escherichia coli* W3101 grown in carrageenan gel beads. FEMS Microbiol Lett 56:307-312

Sayadi S, Nasri M, Barbotin JN, Thomas D (1989) Effect of environmental growth conditions on plasmid stability, plasmid copy number and catechol 2,3-dioxygenase activity in free and immobilized *E.coli* cells. Biotechnol Bioeng 33:801-808

Saye DJ, Ogunseiten O, Sayler GS, Miller RV (1987) Potential for transduction of plasmids in natural freshwater environment: effect of plasmid donor concentration and a natural microbial community on transduction of *Ps aeruginosa*. Appl Environ Microbiol 53:987-995

Sharp R, Bryers JD, JonesWG (1998) Activity and stability of a recombinant plasmid-borne TCE degradative pathway in biofilm cultures. Biotechnol Bioeng 59:318-327

Shu CH, Yang ST (1996) Effect of particle loading on GM-CSF production by *Saccharomyces cerevisiae* in a three-phase fluidized bed bioreactor. Biotechnol Bioeng 51:229-236

Sode K, Morita T, Peterhans A, Meussdoerffer F, Mosbach K, Karube I (1988) Continuous production of α-peptide using recombinant yeasy cells. A model for continuous production of foreign peptide by recombinant yeast. J Biotechnol 8:113-122

Vieth WR (1989) Inducible recombinant cell cultures and bioreactors. In Fiechter A, Okada H, Tanner A, Bioproducts and Bioprocesses, Springer-verlag, Berlin, pp 51-70

Walls EL, Gainer JL (1989) Retention of plasmid bearing cells by immobilization. Biotechnol Bioeng 34:717-724

Walls EL, Gainer JL (1991) Increased protein productivity from immobilized recombinant yeast. Biotechnol Bioeng 37:1029-1036

Wu K, Wood TK (1994) Evaluation of hok/sok killer locus for enhanced plasmid stability. Biotechnol Bioeng 44:912-921

Yang ST, Shu CH (1996) Kinetics and stability of GM-CSF production by recombinant yeast cells immobilized in a fibrous-bed bioreactor. Biotechnol Prog 12:449-456

Zhang Z, Moo-Young M, Christi Y (1996) Plasmid stability in recombinant *Saccharomyces cerevisiae*. Biotechnol Adv 14:401-435

Zhang Z, Sharer J, Moo-Young M (1997a) Protein production using recombinant yeast in an immobilized-cell-film airlift bioreactor. Biotechnol Bioeng 55:241-251

Zhang Z, Scharer J, Moo-Young (1997b) Plasmid instability kinetics in continuous culture of a recombinant *Saccharomyces cerevisiae* in airlift bioreactor. J Biotechnol 55:31-41

Zukowski MN, Gaffney DF, Speck D, Kauffman M, Findelli A, Wisecup A, Lecocq JP (1983) Chromogenic identification of genetic regulator signals in *Bacillus subtilis* based on expression of a cloned *Pseudomonas* gene. Proc Natl Acad Sci USA 80:1101-1105

Immobilization for High-Throughput Screening

NICOLE M. NASBY, TODD. C. PETERSON, and CHRISTOPHER J. SILVA

Introduction

Immobilization is an important and emerging approach in biotechnology for investigation into microbial and cellular processes. Immobilization methods have been used in the production of antibiotics (Ogaki, 1986; Mahmoud et al., 1987) and in other applied fields including, transplantation (Cotton, 1996; Hagihara et al., 1997), clinical (Chang, 1995; Wang and Wu, 1997), food (Larisch et al., 1994) and environmental (Weir et al., 1995; Russo et al., 1996) science. Recently immobilization and bioencapsulation approaches developed at ChromaXome Corporation (Nasby and Peterson, 1998) have combined, adapted and refined a number of previously explored uses of bioencapsulation technology for application in the field of high-throughput screening.

Over the past 6 to 8 years, the drug discovery process has been dramatically impacted by the emergence of three important and complementary technologies: genomics, combinatorial chemistry and high throughput screening. The need for highly efficient screening systems continues to grow today in concert with greater access to novel pharmaceutical targets and available chemistries through development of genomics and combinatorial chemistry technologies. Developmental and applied trends in high-throughput screening over the past several years have steadily moved towards miniaturization and automation of biological assays (Major,

Nicole M. Nasby, Oregon State University , 104 Ocean Administration Building, Corvalis, OR, 97331, USA

Todd. C. Peterson, Genicon Sciences Corporation, Technology Development, 11585 Sorrento Valley Road, San Diego, CA, 92121, USA

✉ Christopher J. Silva, TerraGen Discovery Inc., Suite 300, 2386 East Mall, Vancouver, BC, V6T 1Z3, Canada

(*phone* +01-604-221-8896 ext. 502; *fax* +01-604-221-8881;

e-mail csilva@terragen.com)

1998). The standard 96-well plate format is being replaced by increasingly smaller volume systems and augmented with new technologies to expand assay capacity at reduced cost.

Immobilization using bioencapsulated systems represents an alternative way of achieving these important high throughput biological assay objectives. Current development efforts employ matrices such as calcium alginate for the immobilization of bioassays. In this assay format large numbers of isolated, independent microenvironment assay units are readily generated in a manner conceptually analogous to microtiter wells. Developed immobilized bioassays are easily manipuable and broadly compatible with a variety of cell based bioassay targets for high throughput screening of microbial natural products and solid-phase chemical libraries. In this chapter we will discuss the advantages, applications and future potential for using immobilization as an important approach for drug discovery and high-throughput screening.

Immobilized bioassay screening formats

The utility of immobilization technologies for high throughput screening is realized through spatial positioning of a compound producer or carrier and a cell-based target system in droplets which comprise calcium alginate matrix assay units. In cases where the compound producer is a natural product producing microorganism, the screening format and droplet formulation must accommodate the specific nutrient and temporal requirements for natural product biosynthesis. This can be achieved using appropriately formulated alginate droplets in a variety of assay configurations involving either direct coencapsulation, double layer encapsulation or exposure of immobilized producer to a target cell lawn (Figure 1).

As diagramatically represented in Figure 1, coencapsulation and double encapsulation assay formats immobilize the compound producer and target cells in the same alginate micro-environment unit, however there are conceptual and practical differences between the formats. Coencapsulation is operationally the most simple immobilized bioassay format to establish, however it is best suited for producer and target organisms with similar growth rates and compatible nutrient and temperature requirements. In this assay format (Figure 2), coencapsulated target cells are killed or inhibited by bioactive compound(s) derived from the producer during the assay period. Assays are generally scored using direct visual discrimination of negative, opaque droplets resulting from dense target cell growth and positive, transparent droplets in which target cell growth is inhibited. Alternatively, encapsulation assays can be established to provide more specific information about the nature of the biological activity beyond target cell growth inhibition. This can be done by utilizing target cells engineered with promoter/reporter gene systems that display a detectable signal in response to particular physiological response (Mamber et al., 1986).

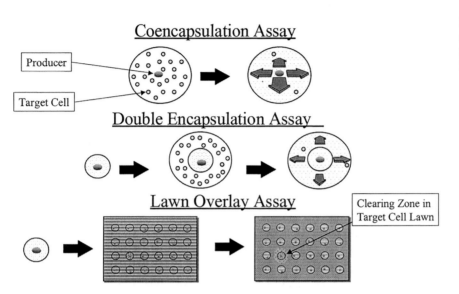

Fig. 1. Macro-droplet Encapsulation Assay Formats (see text)

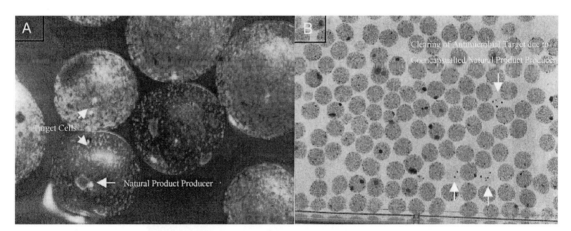

Fig. 2. Coencapsulation of antimicrobial target cells with two *Streptomyces* species. Opaque macrodroplets contain control *Streptomyces* strain that lacks anti-microbial activity. Transparent macrodroplets (arrows) contain *Streptomyces* strain that produces an active natural product. Close-up (**A**) and over view (**B**) images of target growth inhibition due to the antimicrobial activity of produced natural product are shown

Double encapsulation of producer and target organisms (Figure 3) provides a bioassay format that offers considerable operational flexibility as distinct nutrient and temperature conditions between the compound producer and target cells can generally be accommodated. Assays in this format procede in two phases where the encapsulated producer is initially cultured under conditions that promote biosynthesis and accumulation of the bioactive compound within the droplet. At this point, the droplets are processed through a brief wash step and a second layer of alginate harboring the target cells is added. Accumulated bioactive compound rapidly diffuses into the thin outer layer to affect growth or other biological properties of the target cells. Double layer encapsulation assays can be scored by noting overt outer layer visual differences between positives and negatives in the manner described above.

The target lawn immobilized cell assay format (Figure 5) is analogous to the double layer format as production and accumulation of bioactive compound(s) occur in the initial phase. Upon the completion of the intitial phase, the droplets are exposed to target cells or organisms contained in a lawn overlay. Accumulated compounds produced in the droplets passively diffuse into the surrounding external environment containing the assay target. Droplets containing producers with bioactive chemical(s) are identified by an obvious zone of clearing or other physiological signal surrounding the positive droplet. In contrast to microencapsulation methods previously described for biological screening purposes (Weaver 1986;

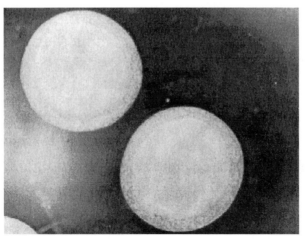

Fig. 3. Double encapsulation antimicrobial assay format with a *Streptomyces* producer (inner core) and target cells (outer layer)

Weaver et al., 1988), the formats, dimensions, configurations and formulations above offer greater developmental flexibility. These approaches are particularly useful for high throughput screening of large numbers of natural or recombinant microorganisms across a range of target cells and organisms with diverse biological, morphological and physiological properties.

Encapsulation of cells is achieved by alginate gelation (Smidsrod et al. 1990) using standard methods (Jekel 1996) for the encapsulation of a single species or the coencapsulation of multiple species (O'Reilly et al. 1995). Double encapsulation is a form of coencapsulation achieved through coating with a second layer of alginate by gently mixing droplets in an alginate bath (Vorlop et al. 1987) containing target cells or organisms. Primary work has been done to develop this technology using microbial growth inhibition assays. In addition, encapsulation bioassays based upon promoter induction of reporter gene(s) expression have been developed. Early target cells for encapsulation bioassays have represented a wide variety of bacteria and fungi including many human and plant pathogens.

High-throughput screening for natural products and combinatorial biology

Natural products screening

Many important pharmaceutical products on the market today are natural products or are chemically derived from natural products isolated after microbial fermentation. Historically, actinomycetes are one of the most important groups of microorganisms in drug discovery due to their ability to produce a wide variety of structurally diverse natural product compounds with a range of important biological activities (Baumberg et al. 1991). Actinomycetes are Gram$^+$, filamentous bacteria that have a sophisticated life cycle characterized by a number of highly regulated and environmentally influenced processes. Actinomycete cultures undergo morphological development, biochemical differentiation and the production of secondary metabolites or natural products that are not essential for the vegetative growth of the organism. These secondary metabolites are generally believed to play important survival, defense or communication functions in nature.

Traditionally, natural product drug discovery from microbes has generally been a laborious and costly process that involves collection, cultivation and fermentation of bacteria and fungi from various environments under various conditions. Determination of biological activity and chemical structure requires an often arduous effort of chemical extraction, analysis of complex chemical mixtures and compound isolation. The use of an immobilization approach streamlines the initial fermentation, extraction, and bioactivity screening aspects of this process into a single step. In addition, immobilization approaches allow for high throughput screening of large numbers of microorganisms under different sets of culture conditions to simultaneously establish parameters for production of natural products of potential interest and to rapidly identify and isolate microorganisms producing natural products with desired biological activities.

In practice, actinomycete cells in the form of spores, protoplasts or mycelial fragments can be encapsulated and cultivated in formulated calcium alginate droplets to produce natural products and spatially separate large numbers of cells into independent micro-environments. Once the cells are immobilized, the alginate droplets are transferred to sterile culture trays and incubated under appropriate conditions for a number of days. Significantly, natural products produced under such conditions in alginate droplets during this culture period accumulate and are retained within the droplets. This process differs significantly from other immobilization production approaches whereby the desired product passively diffuses from the immobilized cells into the surrounding growth medium prior to isolation. Importantly, the process described above allows one to directly expose the droplets to a biological assay at an appropriate stage of development for natural product production and recover the viable producing organism after screening. For screening of natural products, this approach obviates the need for large-scale independent liquid cultivation and chemical extraction of individual microbial isolates prior to biological screening. Given the progress to date using actinomycetes, it is highly probable that these assay systems can be developed to accommodate other important natural product producing microbes with complex lifecycles such as myxobacteria and fungi.

Screening combinatorial biology libraries

Fungi and bacteria, including actinomycetes, have provided many natural products that have become drugs or lead compounds. However, traditional natural products discovery is necessarily limited to cultivable organisms that comprise only an estimated 1-5% of total potential microbial biodiversity present in natural environments. Some cultivable microbes that produce interesting natural products often require special growth conditions for efficient production that are economically or procedurally incompatible with industrial processes. Combinatorial biology is a strategy that applies molecular biology and industrial microbiology to construct DNA expression libraries from biosynthetically talented microbes. In this natural products discovery process (Thompson et al., 1998), DNA from either pooled donor microbes or directly isolated from an environmental sample is prepared and cloned into vectors that facilitate transfer and heterologous expression of metabolic pathways for the biosynthesis of natural products in appropriate, characterized industrial hosts (Figure 4). Results from an initial technology assessment project have demonstrated that structurally novel bioactive compounds can be generated by combinatorial biology (Peterson et al., 1997). Given the large number of independent recombinant expression library clones one can generate using currently developed genetic tools for this technology, immobilization approaches for high throughput screening are a critical component of this discovery system.

For natural products discovery in actinomycetes, alginate encapsulation approaches are particularly useful given the inherent procedural flexibility and the ability of developed assay formats to accommodate the developmental/temporal aspects of natural product biosynthesis and spatial requirements in bioassay design. Overall, the fundamental advantages of this process are its ability to efficiently identify and isolate recombinant clones that produce natural products with a high probability of structural novelty and desired biological activities.

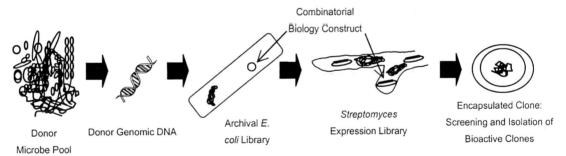

Fig. 4. Integrated Combinatorial Biology Discovery Process: Genomic DNA is isolated from pooled cultivable donor species or environmental samples and archival genomic libraries in *E. coli* are constructed. The archival genomic libraries are then transferred from *E. coli* to an appropriate expression host(s) and recombinant clones are cultured in formatted alginate macrodroplets to promote production and accumulation of natural products. A variety of assay formats are then used to screen expression clones for production of biologically active molecules. Identified clones are then isolated for further chemical and genetic analysis

The encapsulation-based screening format is particularly useful in high throughput screening combinatorial biology libraries that may contain millions of independent recombinant clones. The streamlining of the component steps of genetic manipulation, fermentation and screening into a single integrated process is crucial due to the generation of high number of clones available for screening. Given the largely random nature of the molecular cloning steps in the process and the efficiency of established cloning systems for library generation, libraries comprised of large numbers of recombinant clones are required to ensure that the biosynthetic potential of the sample will be harnessed through adequate representation of the donor genomic DNA. This is a particularly important consideration when generating and screening combinatorial biology libraries from biodiverse and complex environmental samples.

Additional efficiency in the combinatorial biology discovery process can be obtained at the step in which libraries are transferred from the *E. coli* archival library construction host to the expression host system(s) of choice. The encapsulation process for *Streptomyces* expression library clones may begin by preparing recipient cells as either mycelia fragments, spores or protoplasts. Transfer of the library DNA to the expression host can be achieved by direct encapsulation of *Streptomyces* protoplasts that have acquired the library through DNA transformation (Hopwood et al, 1985) or expression host cells that have received the archival library via inter-species conjugation (Mater 1996; Peterson et al., 1998). Expression library clones generated by these manipulations are allowed to grow, develop and produce in macrodroplets composed of nutrient medium in an alginate matrix. For some systems, an additional benefit associated with the ability of the encapsulation environment to promote plasmid stability during the growth phase (Barbotin 1994) and enhance production of natural products may be obtained by this approach. During production, natural products accumulate and concentrate within this environment, and clones with interesting bioactivities can then be screened against a diverse panel of biological targets. Viable, bioactive expression clones can be easily recovered from large combinatorial biology libraries.

The efficiency of this discovery system has been demonstrated in a model system for generating and screening a simple actinomycete expression library. A genomic library of *Streptomyces avellaneus*, a known tetracycline producer, was prepared in *E. coli* and introduced into *Streptomyces lividans*, a heterologous expression host, by DNA transformation. Approximately 12,000 spores prepared from a pool of 2000 recombinant *S. lividans* expression clones were screened in alginate droplets for antimicrobial activity against an *E. coli* target. On statistical average, one or two

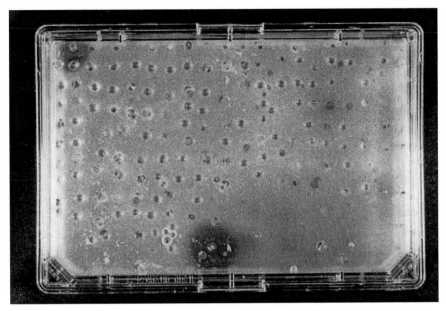

Fig. 5. Macrodroplet Screening of an Actinomycete Expression Library. Screening of clones from a *Streptomyces lividans* expression library for antimicrobial activity using the lawn overlay format. Positive clones identified from encapsulated developed colonies exhibiting a zone of clearing in the lawn of *E. coli* target cells are easily isolated for further chemical and genetic analysis

recombinant spores were encapsulated per macrodroplet and cultured for 14 days. Macrodroplets were overlayed with a lawn of *E. coli* and screened for antimicrobial activity according to established zones of clearing in the target lawn (Figure 5). A total of 18 clones with activity against *E. coli* were identified. These positive clones were then isolated from the target overlay and cultured on nutrient agar plates. The recovered clones were subjected to further chemical and genetic analysis to confirm their activity in anti-microbial bioassays and heterologous expression of the tetracycline bio-synthetic pathway.

Other screening applications and future directions

Target selection and assay development

While the developed growth inhibition and promoter induction assays in microbial systems are useful and informative, future advancement for this approach in high throughput screening will focus on the adaptation of immobilization technologies to develop robust assays with cell-based and molecular target systems that are of pharmaceutical importance. Initial studies have indicated that adaptation of the approach to develop cell-based assays using mammalian and plant targets will be straightforward. The immobilization technologies are also well suited for the development of high throughput biological assays using a wide variety of small, whole organism targets such as nematodes, parasites and insect crop pests. Studies to develop immobilization assay technologies that further broaden the scope and utility of the approach are in progress.

Automation

The encapsulation assay technology is readily amenable for scale-up and automation in order to further increase its utility in high-throughput screening. This critical aspect may be explored through the development of three automation components: a bioencapsulation instrument to generate large numbers of formatted macrodroplets, robotics to automate intermediate handling and incubation steps and a sorter instrument with flexible detection capabilities and separation mechanics to score, collect and process positives. All of these components have either already been independently developed for related applications or require some minimal modification of available instrumentation, robotics, optics or software. Efforts to enhance the utility of immobilization assays for high throughput screening through the development of automation instrumentation are currently underway.

Combinatorial chemistry library screening

This encapsulation bioassay approach can also be applied to the screening of combinatorial chemical libraries. Conceptually, the coencapsulated "producer" in this application is the carrier of a library chemical rather than a natural product producing microorganism. This application may be developed using solid phase chemical synthesis and photo-cleavable linker chemistries (Holmes, 1997). Compounds linked to solid phase beads can be independently encapsulated with desired target cells or organisms. Photo cleavage of library compounds off the beads with an appropriate light pulse allows the compound to diffuse through the matrix and interact with the bioassay target. Once positives are identified, additional com-

pound remaining on the shadow side of the droplet can be recovered for
further analysis.

References

Barbotin, J-N (1994) Immobilization of Recombinant Bacteria. A strategy to improve
plasmid stability. Ann. NY Acad. Sci 721:303-309

Baumberg S, Krugel H, Noack D (ed.) (1991) Genetics and product formation in *Streptomyces*. Plenum Press, New York.

Cotton, CK (1996) Engineering challenges in cell-encapsulation technology. Trends
Biotechnol. 14:158-162.

Chang, TM (1995) Artificial Cells with Emphasis on Bioencapsulation in Biotechnology.
Biotechnol. Annu. Rev. 1:267-295.

Hagihara Y, Saitoh Y, Iwata H, Taki T, Hirano S, Arita N, and Hayakawa T (1997) Trans-
planation of xenogeneic cells screting beta-endorphin for pain treatment: analysis of
the ability of component of complement to penetrate through polymer capsules. Cell
Transplant. 6:527-530.

Holmes, CP (1997) Model Studies for New *o*-Nitrobenzyl Photolabile Linkers: Substi-
tuent Effects on the Rates of Photochemical Cleavage. J. Org. Chem. 62:2370-2380.

Hopwood, DA et al., (1985) Genetic Manipulation of *Streptomyces*, A Laboratory Manual. The John Innes Foundation, Norwich, United Kingdom.

Jekel M, Vorlop K-D (1996) Beioencapsulation technology overview. Procedings from
the 5th International Workshop on Bioencapsulation, Potsdam, Germany.

Larisch, BC, Poncelet, D, Champagne, CP and Neufeld, RJ (1994) Microencapsulation of
Lactococcus lactis subsp. *cremoris*. J. Microencapsul. 11:189-195.

Mamber SW, Okasinksi WG, Pinter CD and Tunac JB (1986) The *Escherichia coli* K-12
SOS chromotest agar spot test for simple, rapid detection of genotoxic agents. Mutat.
Res. 171:83-90.

Mahmoud W and Rehm JH (1987) Chlorotetracycline production with immobilized
Streptomyces aureofaciens. Appl. Microbiol. Biotechnol 26:333-341.

Major J (1998) Challenges and Opportunities in High Throughput Screening: Implications for New Technologies. J. Biomolecular Screening 3:13-17.

Mater DDG, Barbotin J-N, Truffaut N and Thomas JC (1996) Cell co-immobilization
with polysaccharidic gel beads: a model system of bacterial conjugation. Proceedings
from the 5th International Workshop on Bioencapsulation, Potsdam, Germany.

Nasby, NM and Peterson, TC (1998) Methods for Screening Compounds using Encapsulated Cells. PCT Patent Publication Number WO 98/41869.

Ogaki M, Sonomoto K, Hiroki N, Tanaka A (1986) Continuous production of oxyte-
tracycline by immobilized growing *Streptomyces rimosus* cells. Appl. Microbiol.
Biotechnol. 24:6-11

O'Reilly AM and Scott JA (1995) Defined coimmobilization of mixed microorganism
cultures. Enzyme and Microbial Tech. 17:636-646

Peterson, TC, Brian P, Foster LM, Li K, Fielding RJ, Thompson KA, McClure G, Rupar LC, Mamber SW, Brooksire KW, Belval R, Pack E, Gugliotti K and Forenza S (1997) Diverse bioactivites from actinomycete combinatorial biology libraries. In: Proceedings from the 1996 Genetics and Molecular Biology of Industrial Microorganisms Conference. (R. Baltz, G. Hegemen and P. Skatrud, eds.), pp. 71-76. Society For Industrial Microbiology Press, Fairfax, Virginia.

Peterson, TC, Foster, LM and Brian, P (1998) Methods for Generating and Screening Novel Metabolic Pathways. United States Patent Number 5783431.

Russo A, Moenne-Loccoz Y, Fedi S, Higgins P, Fenton A, Dowling DN, O'Regan M and O'Gara F (1996) Improved delivery of biocontrol *Pseudomonas* and their antifungal metabolites using alginate polymers. Appl. Microbiol, Biotechnol. 44:740-745.

Smidsrod O and Skjak-Braek G (1990) Alginate as immobilization matrix for cells. Tibtech. 8:71-78

Thompson, KA, Foster, LM, Peterson, TC, Nasby, NM and Brian, P (1998) Methods for Generating and Screening Novel Metabolic Pathways. United States Patent Number 5824485.

Vorlop K-D, Steinert HJ and Klein J (1987) Cell immobilization within coated alginate beads or hollow fibres formed by ionotropice gelation. Enzyme Eng. 8:339-342

Wang, N and Wu, XS (1997) Preparation and characterization of agarose hydrogel nanoparticles for protein and peptide drug delivery. Pharm. Dev. Technol. 2:135-142.

Weaver JC, (1986) Gel microdroplets for microbial measurement and sccreening: basic princples. Biotechnology and Bioengineering Symp. 17:185-195

Weaver JC, Williams GB, Klibanov A, Demain AL (1988) Gel microdroplets: rapid detection and enumeration of individual microorganisms by their metabolic activity. Bio/Technology 6:1084-1089

Weir, SC, Dupuis, SP, Providenti, MA, Lee, H, and Trevors, JT (1995) Nutrient-enhanced survival of and phenanthrene mineralization by alginate-encapsulated and free *Psuedomonas* sp. UG14Lr cells in creosote-contaminated soil slurries. Appl. Microbiol. Biotechnol. 43:946-951.

Acknowledgements. We wish to acknowledge Nina Aronson, Alex Cantafio, Tina Legler, Heather Elbert, and Jennifer Englehardt for performing much of the work described in this chapter.

Subject Index

A

Actinomycetes 251, 252
Abrasion 36, 184, 193, 195
Agarose 130, 133
Airlift loop reactor 164, 165, 183, 184
Alginate 6, 20, 25, 133, 153, 178, 201, 214–217, 251
Animal cells 124
Attachment 197
Auto fluorescence 111
Automation 256

B

Bacillus 237
Bacteriophage infections 1, 201, 204
Basement membrane gel 129
Batch operation 201, 207
Beer 205
Biodegradability 192
Bioencapsulation 247, 248
Biomass concentration 65–73, 123, 239
Biomass gradients 74
Bioluminesence 230
Bioremediation 213–234
Break up 151
Bubble column 164
Burkholderia cepacia 237

C

Candida tropicalis 126
Capillary jet 151
Carrageenan 6, 23, 67, 69, 103, 133, 152, 167, 201, 217–219, 237, 238
Catechol 2,3-dioxygenase 240
Catharanthus roseus 125
Cathode 87, 89
Cell number 71

Champagne 1
Cheese 1, 200
Chitosan 25
Chlorophenol 213
Coal tar 213
Coating 25
Co-immobilization 219–221, 248
Combinatoorial biology 253
Compression 39
Conservation 205, 207
Contamination 205
Continuous operation 201, 238, 242
Coomassie Brilliant Blue 69, 70
Costs 197
Creep measurement 38, 40
Cryogel 223
Cyanobacteria 95

D

Decay 77
Density 3, 31
Destruction 66
Desulfovibrio gigas 126
Diameter 3, 35
Diffusion cell 45–54
Diffusion coefficient 3, 44–64, 93, 94, 126, 162, 184
Diffusion limitation 74, 77–84, 163, 182, 187, 242
Dynamic models 77
Dunaliella salina 125

E

Elasticity 37
Electrostatic droplet generator 16
Emulsification method 18, 146–148
Energy dissipation 165

Enterobacter agglomerans   204
Environmental biotechnology   213–234
Escherichia coli   204, 236, 237, 242
Ethanol   205
Expression library   254, 255
Extraction   66
External mass transfer   162–174

F
Fatigue   195
Fibroblasts   124
Film theory   163
Fluidization   175
Fluorescent antibodies   116
Fluorescin diacetate   104
Food safety   206
Fracture properties   38, 39
Freeze drying   209

G
Galileo   164
Genetically modified organisms
    1, 235–246
Growth   74, 77, 101, 182, 197, 241
Glucuronic acid   215

H
Hibridomas   124
High fructose syrup   205
High throughput screening   247, 248

I
Image analysis   102, 166
Immobilization technique   15–30, 37,
    150–159
Immobilization equipment   16, 216
Inoculation   202–204, 230, 241
Internal diffusion limitation   77–84
Ion exchange   163, 167
Islets of Langerhans   1

J
Jet cutting method   16, 141

K
Kinetics   74–76
Kjeldahl   66
Kolmogoroff   164, 165

L
Lactic acid bacteria   1, 200, 204
Lactobacillus brevis   204
Lactococcus lactis   236
Large scale   200
Lawn overlay assay   249, 250
Lentikats   143
Lissamine green   104
Lyophilization   217

M
Maintenance   75
Mannuronic acid   215
Mass transfer   74, 162, 184, 185, 187
Mass transfer coefficient   79, 162–174
Mechanical stability   36, 191
Medium composition   205, 207
Micro-algae   68, 125
Microcarriers   130
Microcolony   101, 216
Micro-electrodes   45, 54–62, 85
Micromanipulator   92
Microscope reactor   103, 112
Milk   205
Mixing   184, 188
Models   74, 77
Monod   75, 81

N
Nitrobacter   102
Nitrosomonas   102
NMR   123–138
Non-Newton fluids   155
Nuclear magnetic resonance
    spectroscopy   123–138

O
Organic solvent   153
Orifice   154
Oscilation test   38, 40
Oxygen micro-sensors   87

P
Particle size   3, 32
Pathogenes   202
Pediococcus acidilactici   236
Pediococcus damnosus   204
Pentachlorophenol (PCP)   213

Perfusion system   128
Photosynthesis   95
Pichia stipitis   126
Picoamperemeter   91
Plant cells   134
Plasmid copy number   240
Plasmid stability   235 – 246
Plasmid vector   238
Polyacrylamide   221
Polyacrylamide-hydrazine   221 – 223
Poly aromatic hydrocarbon (PAH)   214
Poly ethylene glycol   68
Polyvinyl alcohol (PVA)   68, 223 – 225
Plasmid stabilization   1
Polyelectrolyte complex membrane   27
Polyethylene glycol   11
Polyvinyl alcohol   11
Polyurethane   11
Pre-fermentation   202
Process interuptions   199
Protein   65 – 73
Pseudomonas fluorescens   126
Pseudomonas putida   236
Psychotrophic organisms   202

R
Rate limiting step   77, 183
Rotating disk   144, 145
Regime analysis   182 – 190
Reological properties   36, 37
Resonance nozzle   16, 150 – 161
Respiration   71, 95
Reynolds   164, 165

S
Sacchoromyces cerevisiae   67, 70, 126, 204, 237
Sake   205
Satellite   159
Scale up   139 – 149, 150 – 151, 183
Schmidt   164
Selection pressure   236
Shape   3, 34
Shear   194

Sherwood   163, 164, 168, 169
Shrinkage   22, 155
Solubility   192
Sonication   68
Soy sauce   205
Staining   69, 102
Staphylococcus aureus   204
Starter cultures   1
Static mixer   147
Stirred tank   164
Stock cultures   203
Storage   206, 208, 209, 217
Strain   40
Streptomyces avellaneus   254
Streptomyces lividans   254, 255
Stress   40
Stroboscope   153
Substrate concentration profiles   4, 44, 74, 78, 92, 93
Superficial gas velocity   165, 167, 187
Support material   6
Synthetic gel   10

T
Tetracycline   254, 255
Thiele modulus   79, 80, 81
Toxicity   200
Transacylation   26

V
Vibration   153
Viscosity   37, 156
Voidage   176
Volume   4, 32

W
Wastewater treatment   224
Wavelength   152
Wine   205

Y
Yeast   204
Yoghurt   200
Young's compression modulus   40